FISH, FISHING AND FISHERIES

SALMON

BIOLOGY, NUTRITION AND CONSUMPTION

Fish, Fishing and Fisheries

Additional books in this series can be found on Nova's website under the Series tab.

Additional e-books in this series can be found on Nova's website under the E-books tab.

FISH, FISHING AND FISHERIES

SALMON

BIOLOGY, NUTRITION AND CONSUMPTION

PATRIZIO LACOPO
AND
MARCO RIEMMA
EDITORS

Nova Science Publishers, Inc.
New York

LIBRARY OF CONGRESS CATALOGING-IN-PUBLICATION DATA

Salmon : biology, nutrition, and consumption / editors, Patrizio Lacopo and Marco Riemma. p. cm.
Includes bibliographical references and index.
ISBN 978-1-60741-287-8 (hardcover : alk. paper)
1. Pacific salmon. 2. Pacific salmon fisheries. I. Lacopo, Patrizio. II. Riemma, Marco.
QL638.S2S2335 2009
333.95'656--dc22
2009045654

Published by Nova Science Publishers, Inc. ✢ New York

CONTENTS

PREFACE

Pacific salmon are among the dominant fish groups and the main consumers of forage resources in the upper layer of the subarctic Pacific. In the last years, the majority of Pacific salmon species in North America and in Asia have experienced an increase in abundance, and their role in marine ecosystem has changed. This book examines the feeding habits and trophic status of the Pacific salmon in different regions of the subarctic Pacific under the influence of changing environmental factors. Moreover, this book deals with the present-day Norwegian regulations of saltwater salmon fisheries and particularly Norway's attempts to harmonize the interests of people and fish. Polycyclic aromatic hydrocarbons (PAH) constitute carcinogenic environmental contaminants such as dioxins. Fishes and particularly Salmonids are very sensitive to the PAH toxicity. The natural and anthropogenic PAH generations lead to an environmental contamination of water, sediments and feed of Salmonids, then to a fish contamination and finally to a human food contamination eating those contaminated fishes. The authors of this book discuss the monitoring of these chemical contaminants in salmon, essential to evaluate the pollution related to human activities and to guarantee the quality of fish as food for human consumption. In addition, the effects of replacing fish oil (FO) with rapeseed oil (RO) in Atlantic salmon post-smolt diets is discussed, and its subsequent effects on liver and muscle fatty acid (FA) composition and growth are described. This book also describes the high-pressure processing of fresh salmon and light preserved products like cold smoked salmon. Its effect on microorganisms, enzymes and organoleptic properties are analyzed, as well as the process parameters, pressure, duration, and temperature affecting microorganisms, structure and color, which all determine the market chances of the product. Moreover, the authors underline the influence of three main factors – super-chilling, dietary lipids and pre-slaughter crowding stress, on Atlantic salmon flesh quality, with a focus on the role of lysosomes and lysosomal enzymes, cathepsins B and L, in muscle structure degradation and flesh quality.

Chapter 1- The aim of the three independent studies mentioned in this chapter was to underline the influence of three main factors (super-chilling, dietary lipids and pre-slaughter crowding stress) on Atlantic salmon (*Salmo salar* L.) flesh quality, with focus on the role of lysosomes and lysosomal enzymes, cathepsins B and L, in muscle structure degradation and flesh quality. The microstructure of the muscle was studied by light microscopy, and the level of muscle degradation was evaluated by the percentage of detachments between myofibres and between myofibres and myocommata, the percentage of myofibre breakages and contracted myofibres, and the presence of intra- and extra-cellular ice crystals. Liquid leakage and fillet texture evaluated instrumentally as fillet firmness were also investigated. Finally, lysosomal cathepsin B and L activities and relative gene expressions were measured; the activity being determined in the lysosomal fraction when the treatment was applied to the salmon *post-mortem* and in the muscle homogenate for total activity when the treatment was applied *pre-mortem*. The effect of *pre-rigor* filleting on muscle structure integrity was also studied.

Atlantic salmon muscle structure and flesh quality were affected by super-chilling, dietary fatty acids and pre-slaughter crowding stress. Total cathepsin B and/or cathepsin L activities and/or gene expressions were significantly negatively correlated to muscle pH, myofibre-myofibre and myofibre-myocommata attachments, and fillet firmness. Cathepsins B and L were in every case influenced by the three main factors studied, acting at the myofibrillar level, but also at the connective tissue level of the salmon muscle. These lysosomal enzymes seemed to play an important role in Atlantic salmon flesh degradation both *pre-* and *post-mortem*, and should be highly considered in salmon production and consumption quality for consumer and farmer satisfaction. Some other proteases like calpains can however not be totally excluded in fish muscle degradation process, even if the evidence of their involvement has to be more investigated.

Chapter 2- Pacific salmon are among the dominant fish groups and the main consumers of forage resources in the upper layer of the subarctic Pacific. In recent years (1990s–2000s), the majority of Pacific salmon species (*Oncorhynchus* spp.) in North America and Asia experienced an increase in abundance and their role in the marine ecosystem has changed. In many publications the authors consider that the high Pacific salmon abundance increases interspecific food competition, and the density-dependent factors during periods of limited food availability are among the main reasons for the change in salmon production characteristics (length, weight, fecundity and increase in population average age).

The generalization of long-term data series (such as information on salmon and nekton feeding, the total quantitative estimates of nekton biomass and food supply for salmon—macroplankton and small nekton—along with consumption values of forage resources by salmon and other nekton species) integrated with Russian surveys carried out in the pelagic zone of the Far Eastern seas and adjacent waters of the northwestern Pacific Ocean made it possible to formulate a new perception of Pacific salmon status and their role in North Pacific pelagic communities.

Despite the fact that total Pacific salmon abundance increased significantly (concurrently with a decrease in large-size plankton biomass) in the Far Eastern marine ecosystems, there were quite a few cases in which it has impacted diet composition and size and interspecific food competition. Apparently, this is explained by the wide food plasticity of salmon (especially pink and chum salmon as the most abundant consumers), large standing stock of food resources, and some decrease in total nekton abundance.

In addition, Pacific salmon consume much less of the forage resources compared to many major species of nekton. For example, in the Pacific waters of the Kuril Islands, the consumption of forage resources by nekton (excluding Pacific salmon) during the summer period of 2004–2007 was 3.4–4.5 million tons, whereas Pacific salmon utilized 0.5–0.8 million tons. In the southern Okhotsk Sea in 1991 (in the period of high abundance of Pacific sardine *Sardinops melanosticta*), nekton species consumed 2.3 million tons of forage resources during summer, whereas Pacific salmon consumed as little as 0.5 million tons. In 2002–2003 (during the period of high abundance of juvenile walleye pollack *Theragra chalcogramma*), in the upper epipelagic layer of the western Bering Sea the overall amount of forage resources consumed by nekton species (excluding Pacific salmon) during autumn was 8.4–8.6 million tons, but Pacific salmon utilized as little as 1.0–1.3 million tons. In 2004–2006 (during the period of low walleye pollack abundance), the consumption of forage resources by nekton species dropped by 6–10 times. Probably such significant changes in abundance of the main consumers of forage resources (in this case walleye pollack and Pacific sardine) can change the food habits of Pacific salmon. The abundance of Pacific salmon has not increased significantly in comparison with most abundant nekton species and, therefore, Pacific salmon have not affected the normal functioning of pelagic communities.

Generally speaking, it is necessary to note that the Pacific salmon carrying capacity varies in different regions of the subarctic Pacific under the influence changing environmental factors.

Chapter 3- New management advice for salmon stocks in Norway states that full recruitment should be a management target. This will impact the sea-based fishery for salmon. While the river fisheries can be opened or closed based on investigations of the particular river stock condition, a catch at sea may hold salmon from many rivers. The sea fishery thus involves the risk of fishing salmon from one or more stocks below full recruitment, and the new management advice implies that saltwater salmon fisheries be severely reduced and, in some areas, phased out in the years to come. This implies a cultural change, as the salmon fishery goes far back in time in Norway. River salmon was traded as dried or salted as early as the 1500s, and from the 1800s and onwards saltwater salmon fishing was expanded to the outer areas of the fjords. This chapter deals with present day Norwegian regulations of saltwater salmon fisheries and particularly Norway's attempts to harmonize the interests of people and fish. I use cultural and statistical information from the northernmost county in Norway, Finnmark. This is where the major catches of salmon take place. As it is also the area where the major part of Norway's indigenous population is living, Finnmark is a case in which important stakeholders are well represented.

Chapter 4- In the present study, fish oil (FO) was replaced by rapeseed oil (RO) in Atlantic salmon post-smolt diets, and the effects on liver and muscle fatty acid (FA) composition, selected lipogenic and lipolytic enzyme activities and flesh oxidative stability assessed. Fish (220 ± 12g initial BW) were fed five experimental diets for 12 weeks in which the FO was serially replaced by 25, 50, 75, and 100% RO. There were no effects of diet on final weight and biometric indices ($P = 0.118$). The percentage of neutral lipids (NL) linearly decreased in muscle ($P = 0.005$) and increased in liver ($P = 0.001$) with RO inclusion levels. Dietary RO affected the FA composition of muscle and liver NL and polar lipids (PL). In muscle NL, a linear accumulation of total n-6 PUFA was observed with increasing levels of RO ($P < 0.001$), with n-3 HUFA($P < 0.001$) concentrations decreasing in a linear and quadratic pattern. In liver NL, the concentration of total n-6 PUFA accumulated in a linear and quadratic fashion ($P < 0.001$) while n-3 HUFA linearly decreased ($P < 0.001$) with increasing levels of RO. Total n-3 and n-6 PUFA content in the PL fraction were also affected ($P < 0.01$) by dietary input of RO in both liver and muscle. Activity of carnitine palmitoyltransferase I was higher in the liver of fish fed with 100% RO ($P < 0.001$). A time effect ($P < 0.001$) was observed in the concentration of lipid peroxidation products, expressed as thiobarbituric acid reactive substances (TBARS), in fish flesh stored under refrigeration for nine days. Moreover, the progressive inclusion of RO in the feed affected ($P < 0.001$) flesh oxidation stability. In summary, RO can totally replace FO in Atlantic salmon feed producing a decrease in flesh susceptibility to lipid oxidation with no deleterious effect on growth performance. However, a 100% RO diet increased ($P < 0.001$) total n-6 PUFA concentrations and decreased ($P < 0.001$) the concentrations of eicosapentaenoic and docosahexaenoic acids in the fillet.

Chapter 5- Polycyclic aromatic hydrocarbons (PAHs) constitute carcinogenic environmental contaminants such as dioxins. They are found in all biotopes of living organisms. Fish, particularly salmonids, are very sensitive to PAH toxicity. The natural and anthropogenic generation of PAHs lead to environmental contamination of water, sediments and feed of salmonids, then to fish contamination and finally to human food contamination when the affected fish are eaten. Therefore, they are responsible for an initial food contamination because they are present in raw food material. The monitoring of these chemical contaminants in salmon is essential to evaluate pollution related to human activities and to guarantee the quality of fish as food for human consumption.

Chapter 6- Important changes in food consumption and distribution systems during the last 50 years have been related to the industrialization of food processing. Salmon, initially produced in salt-dried, cold-smoked form by simple tradesmen, was originally regarded as a luxury product, which has now become an industrial product (Cardinal, 2001). While the introduction of aquaculture of the Atlantic salmon has increased the consumption of smoked salmon, problems related to quality still persist. Even there does not exist standardized criteria for the shelf-life, the commercial shelf-life is usually limited to only one week when chill-stored.

Non-thermal food preservation in particular high-pressure processing becomes more applied in the industries. Prolonging of the shelf life and improvement of the safety of the products are the key issues. High-pressure processing of fresh salmon and light preserved products like cold smoked salmon are reported; the effect on microorganisms, enzymes, and organoleptic properties are described. First, the general effects of high-pressure on food constituents mainly proteins are discussed. The secondary, tertiary and quaternary structure of proteins can unfold due to the effect of high pressure, resulting in the inactivation of enzymes. Enzymes contribute to the softening of salmon meat, off-taste and off-odors during storage. Lipids and proteins/enzymes are constitutes of the cell membranes. It is known that high-pressure can inactivate microorganisms by damaging their membranes. A further topic is related to the effect of high-pressure on the color. A bleaching of salmon meat after high-pressure processing is often observed. The color is mainly caused by the pigment astaxanthin. Due to the antioxidative properties of astaxanthin lipid oxidation and color should be related, as antioxidants can trap free radicals, preventing lipid oxidation. The process parameters, pressure, duration, and temperature affecting microorganisms, structure, and color are described in this chapter. All these effects have to be evaluated as they determine the market chances of the product.

Chapter 7- In response to protectionist pressures from United Kingdom salmon producers, in 1996 Norway instituted a feed quota to limit its production. The feed quota made the Norwegian and world supply curves for salmon less price elastic. This contributed to price instability, magnified the downward pressure on world salmon prices associated with the 71% increase in supplies of farmed salmon from Chile between 2000 and 2002, and increased Norway's incidence of US salmon tariffs. Our analysis suggests these unintended consequences of the feed quota policy are nontrivial, amounting to some $62 million in lost surplus to UK producers between 2000 and 2002, equivalent to 9.5% of export value.

Chapter 8- Agricultural industrialization and globalization are often identified as key processes in the erosion of food security and sovereignty. Global seafood production is dominated by modern day hunter-gatherers and so food sovereignty debates until recently have been confined to terrestrial systems. The recent introduction and subsequent spectacular proliferation of industrial marine aquaculture has brought the food sovereignty debate to the sea. The current hub of North American marine-based industrial aquaculture is British Columbia (BC) where salmon farming has grown to become a major factor in shaping coastal communities. Since its introduction into BC in the mid-1980s, the global industrial salmon farming industry has changed dramatically. The meteoric rise of farm salmon production exported from Norway, Chile, and Scotland has caused global salmon prices to plummet, forcing intense industry-wide rationalization. This is typically characterized by large multinationals cannibalizing smaller producers in order to establish economies of scale to counterbalance market price erosion of their product. As a result, global market forces have all but eliminated the sovereignty of both BC salmon farmers and salmon fishermen. Salmon farmers and fishermen—and their communities—must adopt ever-greater economies of scale to remain competitive with growing/consolidating foreign farm producers. All ecological issues associated with farms (organic wastes, parasites, toxins use, escapes etc.) are density dependant and therefore escalation of production results in incrementally greater ecological impacts. Further, the (d-)evolution of salmon from a culturally iconic seasonal delicacy and cultural keystone species of indigenous communities to a low-value global commodity has manifested a broad array of impacts highlighting the codependence of ecology, culture, and economics in coastal communities. Finally, increasing production increases supply, which in turn lowers price and alters consumer expectations of price/value relationships. This not only forces another round of production increases/consolidation on the farm, but also drives out commercial fishermen, particularly smaller owner/operators, who have no capacity to adopt economies of scale in the face of falling salmon prices. The ensuing rationalization process results in corporate fleets, the only model capable of adapting to the economic realities of the commoditization of salmon. Such loss of sovereignty effectively liquidates social capital of coastal communities which is instead invested into corporate machinery paying a lower rate of return far removed from the communities who are now left to pay off the interest as a legacy to the venture.

> He who has bread has many problems. He who has no bread has only one problem.
> —*Byzantine proverb*

In: Salmon: Biology, Nutrition and Consumption
Editors: P. Lacopo et al., pp. 1-60
ISBN: 978-1-60741-287-8

Chapter 1

ATLANTIC SALMON (*SALMO SALAR* L.) FLESH QUALITY ROLE OF LYSOSOMAL CATHEPSINS B AND L IN MUSCLE DEGRADATION -

D. Bahuaud [1]***, T. Mørkøre*** [21]***, T K Østbye*** [2]***, B.E. Torstensen*** [3]***, E. Veiseth-Kent*** [4]***, B. Ruyter*** [21]***,***
R. Ofstad [4] ***and M.S. Thomassen*** [1]

[1]Department of Animal and Aquacultural Sciences (IHA), Norwegian University of Life Sciences (UMB), Po Box 5003, 1432 Ås, Norway; 2) Nofima Marin, Po Box 5010, 1432 Ås, Norway; 3) National Institute of Nutrition and Seafood Research (NIFES), Po Box 2029, Nordnes, 5917 Bergen, Norway; 4) Nofima Mat AS, Osloveien 1, 1430 Ås, Norway.

ABSTRACT

The aim of the three independent studies mentioned in this chapter was to underline the influence of three main factors (super-chilling, dietary lipids and pre-slaughter crowding stress) on Atlantic salmon (*Salmo salar* L.) flesh quality, with focus on the role of lysosomes and lysosomal enzymes, cathepsins B and L, in muscle structure degradation and flesh quality. The microstructure of the muscle was studied by light microscopy, and the level of muscle degradation was evaluated by the percentage of detachments between myofibres and between myofibres and myocommata, the percentage of myofibre breakages and contracted myofibres, and the presence of intra- and extra-cellular ice crystals. Liquid leakage and fillet texture evaluated instrumentally as fillet firmness were also investigated. Finally, lysosomal cathepsin B and L activities and relative gene expressions were measured; the activity being determined in the lysosomal fraction when the treatment was applied to the salmon *post-mortem* and in the muscle homogenate for total activity when the treatment was applied *pre-mortem*. The effect of *pre-rigor* filleting on muscle structure integrity was also studied.

Atlantic salmon muscle structure and flesh quality were affected by super-chilling, dietary fatty acids and pre-slaughter crowding stress. Total cathepsin B and/or cathepsin L activities and/or gene expressions were significantly negatively correlated to muscle pH, myofibre-myofibre and myofibre-myocommata attachments, and fillet firmness.

Cathepsins B and L were in every case influenced by the three main factors studied, acting at the myofibrillar level, but also at the connective tissue level of the salmon muscle. These lysosomal enzymes seemed to play an important role in Atlantic salmon flesh degradation both *pre-* and *post-mortem*, and should be highly considered in salmon production and consumption quality for consumer and farmer satisfaction. Some other proteases like calpains can however not be totally excluded in fish muscle degradation process, even if the evidence of their involvement has to be more investigated.

1. Introduction and Background

1.1. Aquaculture: Production of Atlantic Salmon

Started in the 1960's, the global production of farmed Atlantic salmon increased from 1,418,000 tonnes in 2007 to 1,479,000 tonnes in 2008. Norway represents the biggest producer of Atlantic salmon (Figure 1), accounting for 51 % of the global production, followed by Chile with 25 % (Eksportutvalget for fisk, 2009).

The increased production of farmed fish, and especially Atlantic salmon, has raised several problems and questions that imperatively have to be solved by the future generations, in order to always provide consumers with a high quality product, using a sustainable production system.

Figure 1. Atlantic salmon farming, Averøy, Norway. Photo: Diane Bahuaud.

1.2. Atlantic Salmon Muscle Structure

Fish flesh contains 66-84% water, 15-24 % protein, 0.1-22 % fat, 1-3 % carbohydrate and 0.8-2 % minerals (Suzuki, 1981). The flesh of teleost fish is divided into blocks of muscle cells separated into myotomes by connective tissue called myocommata (Figure 2). Each muscle cell is surrounded by connective tissue which attaches to the myocommata by means of fine collagenous fibrils (FAO, 2005) (Figure 3).

Skeletal muscle fibres can be classified into 2 main types: red and white (Figure 2). In fish, the two types of fibres are separated from çach other. White fibres contract rapidly but cannot support continuous heavy work. The energy for their activity is derived mainly from anaerobic glycolysis. In fish, red muscle fibres alone are used for low speed swimming, red and white fibres together for high swimming speed, and white fibres alone for very high swimming speed thanks to their high speed of contraction and their orientation within the white muscle (Alami-Durante and Rescan, 2003).

Skeletal muscle is composed of long, cylindrical, multinucleated cells (Figure 3). In muscle, the cell membrane is called sarcolemma; the cytoplasm of the cell, sarcoplasm, and the smooth endoplasmic reticulum, sarcoplasmic reticulum. Skeletal muscle is surrounded by an external sheath of dense connective tissue, the epymisium. Each group of muscle fibres (myofibre) is also surrounded by connective tissue called perymisium. A fascicle is a bundle of perymisium-bound fibres and the endomysium represents the layer of connective tissue composed mainly of a basal lamina and reticular fibres surrounding each muscle fibres (Junqueira et al., 1989a).

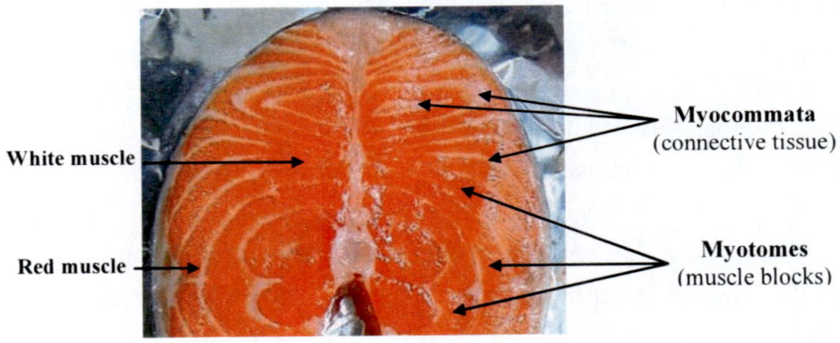

Figure 2. Atlantic salmon muscle Photo: http: //www. earthlife. net/fish /images / anatomy/salmon-ts.jpg.

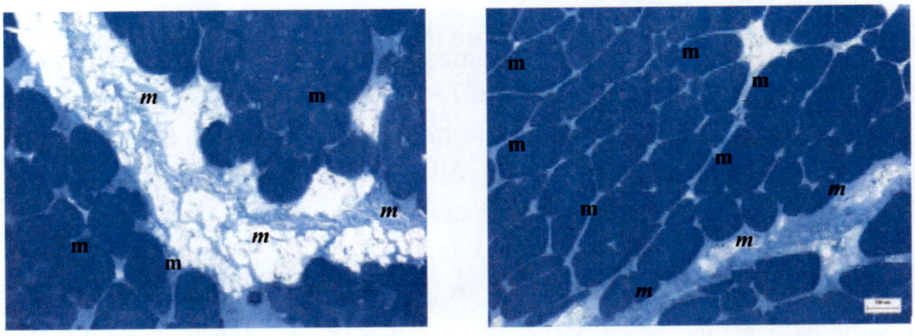

Figure 3. Light microscopy (x20) of Atlantic salmon skeletal muscle. *mc* = myocommata (connective tissue); mf = myofibre (muscle fibre). Scale bar = 100 µm. Photos: Diane Bahuaud.

Myofibres are composed of a bundle of myofibrils. In the skeletal muscle, myofibrils present alternating light and dark bands. The darker bands (A bands) are anisotropic, i.e. birefringent in polarized light; the lighter bands (I bands) are isotropic, i.e. they do not alter polarized light. Each I band is bisected by a dark transverse line, the Z-line. The sarcomere represents the smallest repetitive subunit and extends from Z-line to Z-line. The sarcoplasm contains bundles of myofibrils, which consist of end-to-end chainlike arrangements of sarcomeres, running parallel to the muscle fibre. Thin filaments run between and parallel to thick filaments, and have one end attached to the Z-line. The H band is a zone in the center of the A band consisting of lonely thick filaments and in the M-line, lateral connections are made between adjacent thick filaments.

At least four main proteins compose the striated filaments of the myofibrils: actin, tropomyosin, troponin, and myosin. Myosin and actin together represent 55 % of the total protein of striated muscle and alpha-actinin is a major component of the Z-line. The structure of the myofibrils is kept by alpha-actinin and desmin, which are believed to tie adjacent sarcomeres together (Junqueira et al., 1989a).

For the muscle to be able to contract, myosin needs to bind to actin; this action breaking down ATP rapidly. Releasing a myosin molecule of a thick filament from an actin molecule of a thin filament in order to relax the contraction requires new ATP. After death, when no ATP is available anymore in the muscle, the actin-myosin complex stays bound and becomes stable. This phenomenon is called *rigor mortis*, translating an extreme muscular rigidity (Junqueira et al., 1989a). The time of onset of *rigor mortis* plays a major role in fish muscle *post-mortem* quality (Skjervold et al., 2001c). Pre-slaughter conditions, nutritional status of the fish and storage temperature influence the duration of *rigor*. Entering in *rigor*, the flesh texture toughens and gradually softens again when the fish leaves this state (Rasmussen, 2001).

In the skeletal muscle, the most readily available energy is stored in the form of ATP and phosphocreatine but also in glycogen reserves, constituting about 0.5-1 % of muscle weight. Fatty acids are the main energy source in skeletal muscle for long-term or endurance exercises. On the contrary, during short-term or violent exercises or stress, glucose is metabolized rapidly, mainly from muscle glycogen stores, to lactate, in anaerobic conditions; the oxygen debt being repaid during the recovery period (Junqueira et al., 1989a). In Atlantic salmon farming, pre-slaughter stress leads to a production of lactate, without recovery period, possibly leading to a degradation of muscle quality.

Muscle connective tissue plays an important role in mechanical transmission of contraction forces, and in the case of fish, swimming movements (Junqueira et al., 1989a). Connective tissue provides a matrix that connects and binds the cells, giving support to the body. The major constituent of connective tissue is its extracellular matrix, composed of protein fibres, an amorphous ground substance, and tissue fluid. Embedded in the extracellular matrix are the connective tissue cells (Junqueira et al., 1989b).

- The ground substance acts as a lubricant and a barrier to the penetration in the tissues by foreign particles, and is mainly composed of glycosaminoglycans and structural glycoproteins. The main function of glycosaminoglycans is probably structural, interacting with collagen fibrils to bind these structures together. Glycosaminoglycans are linear polysaccharides and are bound to a protein core, forming a proteoglycan molecule. The degradation of proteoglycans is carried out by several cell types and depends on the presence of several lysosomal enzymes. (Junqueira et al., 1989b).

- There are 3 main types of connective tissue fibres: collagen fibres, reticular fibres and elastic fibres; collagen fibres being the most numerous fibres. Collagen and reticular fibres are formed by the protein collagen, whereas the elastic fibres are composed mainly of the protein elastin. The degradation of collagen is simple and is initiated by specific enzymes called collagenases. These enzymes cut the collagen molecule into 2 parts that are susceptible to further degradation by nonspecific proteases (Junqueira et al., 1989b). Some cathepsins like cathepsin L are also able to degrade native collagen in fish muscle (Yamashita and Konagaya, 1991).
- Cells of connective tissue interact, creating complex mechanisms that help defending the organism from invasion. *Fibroblast* is the most commonly found cell in connective tissue. Fibroblasts synthesize collagen, reticular and elastic fibres, and the glycosaminoglycans and glycoproteins of the amorphous intercellular substance. *Macrophages* generally have a well-developed Golgi complex, many lysosomes, and a prominent rough endoplasmic reticulum. The major function of macrophages is the ingestion of particles and their digestion by the lysosomes and the secretion of substances participating in defensive and reparative functions. Other cells like mast cells, plasma cells, adipose cells, leucocytes are also found in connective tissue (Junqueira et al., 1989b).

Collagen content in fish flesh is below 3 % and appears to be even lower in salmonids' muscle. However, total collagen has been reported to increase the toughness of fish flesh and to influence quality characteristics including tenderness, succulence and elasticity. A high solubility of collagen increases softness of the flesh and the disintegration of collagen increases tenderization (Rasmussen, 2001). Collagenases are enzymes capable of degrading the polypeptide backbone of native collagen. There are two types of proteases with collagenolytic activity: serine collagenase, which probably has a digestive function in fish, and metallocollagenases which cleaves the collagen helix at a specific locus under physiological conditions and are involved in the modeling of the extra-cellular matrix (Kim et al., 2002; Sellers and Murphy, 1981). It is thus possible that collagenases act directly on the degradation of collagenous fibrils in fish muscle (FAO, 2005) and influence fillet texture (Hultmann, 2003; Hultmann and Rustad, 2007).

At least three proteolytic systems are implicated in muscle degradation: the proteasome system, calcium dependent proteases (calpains) and lysosomal cathepsins. In fish, cathepsins and calpains seem to be the most involved systems in *post-mortem* muscle softening including degradation of the fine structure of muscle myofibril and intramuscular connective tissue (Ando et al., 2001; Aoki and Ueno, 1997; Chéret et al., 2007; Okitani et al., 1980; Yamashita and Konagaya, 1991).

1.3. Muscle Lysosomes and Lysosomal Cathepsins

1.3.1. Muscle lysosomes

Lysosomes were first described by de Duve et al. in 1955 and have been considered for many years as garbage compartments of the cell. It is now established that these organelles and their components can play an important role in several biological processes in eukaryotes such as bone resorption, tumor progression and programmed cell death. In mammals, lysosomes are present in all cell types except red blood cells (Tardy et al., 2006).

Lysosomes are organelles considered as the "stomachs" of the cells, because they are the major component of the degradation system. They are composed of a single membrane surrounding a lumen containing a high concentration of numerous digestive enzymes: proteases, glycosidases, lipases, nucleases and phosphatases (Béchet et al., 2005).

Lysosomal hydrolases are physically isolated from cytoplasmic constituents and a breakage of the lysosomal membrane is needed for them to exert their hydrolytic activity (Béchet et al., 2005; Köhler, 1989 and 1991; Tardy et al., 2006). The release of lysosomal proteases into the cytosol may precede mitochondria dysfunction (Tardy et al., 2006). Lysosomal integrity is therefore an indicator of cellular well-being and health (Moore et al., 2006) and the regulation of the lysosomal system and its acid hydrolases is vital for cellular homeostasis (Béchet et al., 2005). The presence of ATP-proton pumps in the lysosomal membrane gives an acidic pH (4.0 – 4.5) to the lysosomal lumen. This acidity favors the digestive activities of the lysosomal hydrolases, and partly unfolds protein-substrates (Béchet et al., 2005).

Two main pathways exist in order to deliver intracellular protein substrates to lysosomes: autophagy (*macroautophagy*, *microautophagy* and *chaperone-mediated autophagy*) and endocytosis (*phagocytosis*, *pinocytosis* and *receptor-mediated endocytosis*) (Béchet et al., 2005; Tardy et al., 2006). The *autophagic process* is implicated in the degradation of cytoplasmic constituents and is generally activated in response to stress conditions. Three types of autophagy can be distinguished (Tardy et al., 2006): (i) in *macroautophagy*, cytoplasmic constituents are sequestered by a pre-autophagosomal structure to generate a vesicle called the autophagosome. The autophagosome then fuses with a lysosome and become an autolysosome, where sequestered constituents are degraded (Béchet et al., 2005), (ii) *microautophagy* consists in the sequestration of cytosolic components directly at the surface of the degradative organelle, and (iii) *chaperone-mediated autophagy* targets to the lysosomal membrane substrate proteins. The substrates for this lysosomal proteolysis are cytoplasmic proteins showing sequences related to the pentapeptide Lys-Phe-Glu-Arg-Gln (KFERQ), which is recognised by a cytosolic molecular chaperone. The chaperone and its substrate protein then bind to the lysosome-associated membrane protein (LAMP-2a) and a translocation of the substrate protein to the lysosomal lumen follows (Béchet et al., 2005; Tardy et al., 2006).

Endocytosis provides lysosomes with extracellular material for digestion. Three distinct processes of endocytosis are known: (i) *phagocytosis* results in the digestion of material such as bacteria and occurs only in certain specialised cells like neutrophils and macrophages, (ii) *pinocytosis* allows for internalisation of soluble material, and in (iii) *receptor-mediated endocytosis*, the recognition of a molecule by its membrane receptor is required to lead to its engulfment (Tardy et al., 2006).

Köhler (1989 and 1991) showed the role of detoxification of the lysosomal system in liver of fish subjected to environmental contamination. Indeed, environmental stress was associated with lysosomal membrane fragility and increase of the number of lysosomes in the liver; this increase in lysosomes logically paralleling an increase in lysosomal enzymes in the cell.

1.3.2. Cathepsins

The main class of lysosomal proteases is represented by the cathepsins, which have been identified in several species, including human, rat, mouse, bovine, pig, chicken and fish. Cathepsins are divided in three groups according to the amino acid of their active site conferring their catalytic activity: cysteine (cathepsins B, C, F, H, K ,L, N, O, S, T, U, W and X), aspartyl (cathepsins D and E) and serine (cathepsins A and G) (Tardy et al., 2006).

Cathepsins are synthesized as pre-pro-enzymes. The pre-peptide brings the cathepsin precursor into the rough endoplasmic reticulum lumen and is then hydrolysed. The pro-peptide enables a proper folding of the nascent cathepsin. In the Golgi network, a phosphorylation of mannose residues occurs, which is an important step for the trafficking of pro-cathepsins to endosomes/lysosomes. In the trans-Golgi network, pro-cathepsins bind to mannose-6-phosphate receptors (M6PR). M6PR and their ligand are then translocated to late endosomes, where pro-cathepsins are dissociated from M6PR by the mildly acidic pH (6.0 – 6.5). The receptors are then recycled back to the Golgi apparatus.

The proteolytic processing of pro-cathepsins into active proteases happens in the late endosomes or in the lysosomes. Late endosomes and lysosomes localise especially around myonuclei, beneath the sarcolemma, and in the interior of the myofibres. The Endoplasmic reticulum (ER) and the Golgi network are the centres of synthesis and trafficking of lysosomal enzymes. They can be found throughout the whole myofibre in all muscles (Béchet et al., 2005; Tardy et al., 2006).

Endopeptidases and exopeptidases are responsible for protein hydrolysis in lysosomes. Endopeptidases cleave the inner peptide bonds of proteins, whereas exopeptidases hydrolyse the C- or N-terminal ends. Cathepsins L, B, D and H are the major lysosomal proteases and have an endopeptidic activity. They are expressed in many tissues, whereas the other lysosomal endopeptidases are less distributed and have a tissue-specific function. As endopeptidase activities occur, new C- and N-terminal residues are formed and can serve as substrates for exopeptidases. Endopeptidases thus probably initiate lysosomal proteolysis (Béchet et al., 2005).

In vitro, all lysosomal endopeptidases can degrade purified myofibrillar proteins, but the specificity of the hydrolysis depends of the type of cathepsin. In fish, cathepsins B, D and L are considered as the enzymes playing the most important role in *post-mortem* muscle softening (Ando et al., 2001; Jiang et al., 1992; Ladrat et al., 2003; Yamashita and Konagaya, 1990 and 1991). In migrating chum salmon, muscle tenderisation occurring during storage was due to hydrolysis of the major proteins constructing the muscle structure. This autolysis was noticed mainly in muscles where the activity of cathepsins B and/or L was very high; cathepsin L being considered as the main enzyme responsible for the myofibrillar protein breakdown (Yamashita and Konagaya, 1990 and 1991).

Cathepsin L (EC 3.4.22.15) hardly hydrolyses Z-Arg-Arg-NMec, which is specific for cathepsin B, but chum salmon cathepsin L shows high proteolytic activities towards the synthetic substrate Z-Phe-Arg-NMec and a variety of protein substrates (Kirschke, 1981). The hydrolytic activity of purified chum salmon cathepsin L towards Z-Phe-Arg-NMec as substrate is maximal at pH 5.6 and the enzyme activity is enhanced by thiol-reducing agents such as EDTA, used in cathepsin activity measurements (Yamashita and Konagaya, 1990). **Cathepsin L** exerts a much higher activity than the other lysosomal proteases on the degradation of various proteins, such as azocasein, hemoglobin, collagen, elastin, and myofibrillar proteins (Barrett and Kirschke, 1981). Cathepsin L is the most probable protease responsible for the degradation of myofibrillar proteins in chum salmon muscle, being able to hydrolyse myosin, troponin T and troponin I, alpha-actinin, nebulin and tropomyosin, but not actin (Aoki and Ueno, 1997; Yamashita and Konagaya, 1991). Cathepsin L from chum salmon can also split native collagen from carp connective tissue at the non-helical region (Yamashita and Konagaya, 1991).

In fish and especially chum salmon and mackerel, cathepsin B (EC 3.4.22.1) is also capable of hydrolysing a large range of proteins, especially tissue proteins (Barrett and Kirschke, 1981). Cathepsin B seemed to have a greater activity on denaturated than on native proteins and exhibit less degradation activity on the myofibrillar fragments than cathepsin L. Cathepsin B can slowly hydrolyse connectin and nebulin, but not collagen, myosin, troponin T, troponin I, and tropomyosin, neither degrade actin (Aoki and Ueno, 1997; Yamashita and Konagaya, 1991). The initial proteolysis by cathepsin L and B could be followed by further degradation of peptides by exopeptidic cathepsins (Tardy et al., 2006). It was also suggested that the action of calcium dependant proteases, calpains, could be the first step before proteolysis by other enzymes like lysosomal cathepsins (Delbarre-Ladrat et al., 2004b; Goll et al., 1992).

1.4. Calpains

Calpains are cytosolic calcium dependent proteases mainly involved in *post-mortem* meat tenderization (Koohmaraie, 1992 and 1996), but could also participate in fish fillet softening and degradation (Chéret et al., 2007; Delbarre-Ladrat et al., 2006; Salem et al., 2004; Taylor et al., 1997). The ubiquitous calpain system consists of μ-calpain, m-calpain and the inhibitor calpastatin and has been widely studied in mammalian meat (Chéret et al., 2007; Goll et al., 2003; Koohmaraie, 1992; Raser et al., 1995; Veiseth and Koohmaraie, 2001). Calcium plays an important role in the activation of calpains (Chéret et al. 2007), the terms μ- and m-calpains (micro molar and milli molar respectively) referring to the concentration of Ca^{2+} required for activation of the calpains (μ-calpain: 50 μM Ca^{2+} and m-calpain: 150-1000 μM Ca^{2+}) (Goll et al. 2003; Koohmaraie, 1992). Mammalian muscle calpains are very specific, digesting primarily troponin-T, desmin, titin, and nebulin, but not actin or myosin (Koohmaraie, 1992). Fish calpains are far more active at low temperatures than calpains from mammalians, and show a species-specific rate of cleavage and an ability to digest myosin (FAO, 2005). Calpains tend to be concentrated in the Z-disk of the muscle and therefore the activity of these proteases could result in the degradation of titin, dystrophin and α-actin, and loss of the Z-disk structure (Papa et al., 1997). Since α-actin is known as a key component in the Z-disc of muscle structure and most calpains are active at physiological pH, calpains could be involved in *post-mortem* textural changes in fish (Delbarre-Ladrat et al., 2006). Little work has, however, demonstrated a significant relationship between calpain activity and instrumental texture measurements (FAO, 2005).

1.5. Fish Quality and Some Influencing Factors

Quality can be defined as "the degree to which a set of inherent characteristic fulfills requirements" (Inteaz, ISO 9000:2000). According to the American society for Quality Source, quality can also be described as "a subjective term for which each person has his or her own definition. In technical usage, quality can have two meanings: (i) The characteristics of a product or service that bears on its ability to satisfy stated or implied needs; (ii) A product or service free of deficiencies." Quality is thus the way a product fulfills the demands and expectations of the consumers. Quality can have a multifaceted definition, related to religion, tradition and use, and different markets can have different requirements regarding quality-related characteristics (Kiessling et al., 2007).

As far as food is concerned, the basic quality parameter is that it is safe to eat (Nortvedt et al., 2007). The primary quality of a food product concerns the hygienic (bacterial load, diseases), nutritional (lipids, proteins, vitamins, minerals), sensory (taste, smell, texture, colour), technological (suitability for freezing, filleting, salting, smoking, etc.) and ethical quality (welfare, sustainable production).

If the product is animal, primary quality also deals with biological aspects (species, size, season, health status) and workmanship in handling and production. The total or secondary quality is more related to the perceived quality (subjective-hedonic) and market quality including delivery according to specifications, price, size of the product, packaging, service, information and distribution (Nortvedt et al., 2007).

Figure 4. Fresh Atlantic salmon. Photo: Diane Bahuaud.

Figure 5. Fresh Atlantic salmon fillets. Photo: *http://www.finecooking.com/* assets/ uploads/ posts/5673/ ING-salmon-filet-2_sql.jpg.

Fish quality perception differs for the fish farmer, the industrial processor and the consumer. Freshness is one characteristic often associated with fish quality; therefore correct handling and storage are very important. The first impression of quality is provided by the outer morphology of a fish (Figures 4 and 5).

Muscle integrity can influence the external aspect of the fillet, but also contribute to sensory quality, texture, and have a strong impact on fillet processing and bacterial development rate. So far, quality criteria for farmed salmonids have mainly taken into account the external characteristics of the fish, whereas intrinsic factors such as muscle integrity, flesh composition and fillet texture have received less attention, except for colour and lipid flesh composition (Rasmussen, 2001). Increased knowledge regarding specific quality characteristics of farmed fish would contribute to the production of even higher quality fish.

Quality in fish can be influenced by numerous factors, both *pre-mortem* and *post-mortem*; among them, breeding, feed composition, feeding regimes, handling and pre-slaughter stress, environment, transport, slaughtering method, filleting technique, storage temperature, processing, and packaging are the most important parameters. It is thus possible to control some aspects of quality during the growth phase, but also *pre-* and *post-mortem*. An effective and rapid handling of large numbers of fish in an industrialised process, with a minimum of stress and pain to the fish, is a difficult operation. Combining the demand for quality, welfare and ethics with the demand for economical production is still a great challenge (Kiessling et al., 2007; Nortvedt et al., 2007).

1.5.1. Super-chilling

Refrigeration of gutted fish or fish fillets is an important parameter and has to be controlled in order to deliver high quality and freshness food products to consumers. At present, it takes four days from the date on which fish are packed in Northern Norway until they are delivered for onward distribution in Europe. Very large quantities of ice are transported together with the fish to keep the product chilled and preserve its freshness and quality. This amount of ice represents about 30 % of the net weight transported, the unit cost for transporting this ice being the same as for transporting the salmon themselves (Haugland, 2001). Because of this loss, it was important to find a technique to reduce transportation costs and transport impact on the environment.

Super-chilling is a technique consisting in freezing or partly freezing the internal water of a food product. The fish are stored at temperatures slightly lower than that used for ice, i.e. between -1 °C and -4 °C. At this temperature, part of the water in the fish flesh is frozen, acting as a refrigeration reservoir, and no external ice is needed (Figures 6 and 7) (Haugland, 2001; Nowlan et al., 1974).

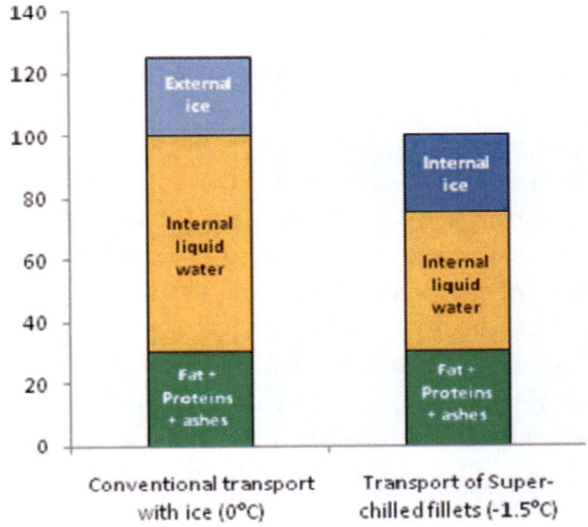

Figure 6. Super-chilling concept. Total transported weight related to transported product (%) (from Haugland, 2001).

Figure 7. Super-chilled Atlantic salmon fillet (frozen upper layer and fresh core) Photo: Norconserv. Forskningsrådet, Nytt fra HAVBRUK nr. 3. Nov. 2007.

The demand for high quality fresh fish products is growing, and exports of fresh fillets have increased in recent years. Super-chilling has already been studied on different food products, most of them emphasizing the many advantages of the technique especially when it comes to freshness and extension of shelf-life (Dion, 1997; Dunn and Rustad, 2007 and 2008; Fik et al., 1988; Gallart-Jornet et al., 2007; Lee and Toledo, 1984; Nowlan et al., 1974; Sivertsvik et al., 2003). However, most of these studies dealt with several days super-chilling storage of gutted fish or fillets processed *post-rigor*. To our knowledge, we reported for the first time a study focusing on a short-term super-chilling process of Atlantic salmon *pre-rigor* fillets, prior to a storage period on ice in a cooling room (Bahuaud et al., 2008).

To a production point of view, super- chilling can increase the yield product, prolong the storage life and freshness of good quality fillets by delaying bacterial spoilage and make the product easier to handle for future processing because of its increased rigidity (Dion, 1997; Dunn and Rustad, 2007 and 2008; Fik et al., 1988; Gallart-Jornet et al., 2007; Haugland, 2001; Hemmingsen and Nordtvedt, 2007; Lee and Toledo, 1984; Nowlan et al., 1974; Sivertsvik et al., 2003). Super-chilling can also make it possible to reach the inland markets, so far considered as out of reach for the fresh fish market industry. More fish could therefore be sold as "fresh" products instead of frozen; giving an added market value to the product (Jensen, 2007).

Trained consumer panels found better quality attributes to super-chilled fish, when scoring on visual aspect and taste (Jensen, 2007). Muscle integrity of Atlantic salmon after super-chilling had however been given little attention, but in a pilot study, Hemmingsen and Nordvedt (2007) noticed that some protein denaturation could happen. Indeed, formation of ice within tissues during freezing can cause irreversible damages. Ice crystallisation occurs in two steps: first, nuclei are formed and second, they grow to a specific crystal size. Freezing rate and final temperature of the process determine the size and location of ice crystals formed during freezing. Slow freezing results in a low number of large ice crystals, while more rapid freezing gives a large number of small ice crystals (Bello et al., 1982; Martino et al., 1998). The challenge of the technique is therefore the optimization and the control of the temperature and the ice-fraction in the product. Formation of ice crystals in the fillet could considerably affect the quality of the final product (Haugland et al., 2006).

1.5.2. Modified atmosphere packaging (MAP)

Modified Atmosphere Packaging (MAP) is a packaging method which modifies the gaseous environment surrounding a food product, in order to extend the shelf-life of this product. The most common gases used in MAP for fish are: carbon dioxide (CO_2), which inhibits bacterial growth, oxygen (O_2), which inhibits the development of trimethyl amine (TMA); and nitrogen (N_2), which is used to prevent package collapsing. In MAP, the initial gas mixture changes, because of the absorption of the gases (mainly CO_2) into the fish product (Hansen, 2008).

MAP is thus used to reduce microbial growth and delay the production of amine compounds (Sivertsvik et al., 2002) of food packed products, even on frozen and thawed fish (Bøknæs et al., 2000). MAP of super-chilled stored Atlantic salmon *post-rigor* fillets has been studied by Sivertsvik et al. (2003). The combination of super-chilling storage with MAP increased the sensory and microbial shelf-life of the fillets by 17 days, when compared to air-stored fillets at chilled temperature.

1.5.3. Pre-rigor filleting

After death, the reserves of ATP in the muscle decrease, and no energy is left for the muscle to relax, leading to the state of *rigor mortis*. In Atlantic salmon production, the time from slaughter to the onset of *rigor mortis* is critical to obtain high quality products. The onset of *rigor* and the rate of contraction can be affected by several factors, including pre-slaughter stress, feed, storage temperature, live-chilling and handling during the slaughtering process (Erikson et al., 1997a and 1997b; Kiessling et al., 2006 and 2007; Mørkøre, 2006; Okamoto et al., 2006; Skjervold et al., 1999 and 2001a; Suontama et al., 2007).

In 2008, 77 % of the Atlantic salmon produced in Norway were exported as fresh gutted fish, but only 9 % as fresh fillets (Eksportutvalget for fisk, 2008). In "normal" conditions (minimum handling stress and cool temperature storage), Atlantic salmon turn into *rigor* between 8 and 70 h after slaughter (Wang et al., 1998). Today, most of the farmed salmon are filleted *post-rigor*, after 3-5 days of storage on ice; this period of storage being needed to reach a total resolution of *rigor mortis*, and then allow an easier removal of pin bones and to avoid processing while the fish is still in *rigor* (Einen et al., 2002a).

On the contrary, *pre-rigor* filleting consists in filleting the fish immediately *post-mortem*, before the onset of *rigor-mortis*. This new concept of producing super-fresh farmed fish fillets has gained much interest in the industry recently. *Pre-rigor* fillets are processed directly after slaughter, so that no storage period before filleting is necessary, allowing the distribution to start earlier on the European market, and shelf-life of high quality fillets to be extended (Nortvedt et al., 2007). Several studies have already shown the great benefits of this process, offering a better freshness, a better colour, a firmer texture, a thicker fillet and less gaping early in the course of storage (Kiessling et al., 2007; Skjervold et al., 2001b and 2001c). *Pre-rigor* filleting also means that other parts of the fish than the fillets themselves become available (Kiessling et al., 2007). *Pre-rigor* filleting increases the value of the fish as fresh fillets. Transport of waste products being reduced by 20 %, this concept can considerably decrease transportation costs and the impact of transport on the environment.

1.5.4. Dietary lipids

Fish quality can be affected by feed composition and different feeding regimes. In Atlantic salmon, the fatty acid composition of the muscle usually reflects the fatty acid composition of the diet (Bell et al., 2003; Thomassen and Røsjø, 1989; Torstensen et al., 2004 and 2005). Atlantic salmon has a low ability to convert linolenic acid (18:3 n-3) into eicosapentaenoic acid (EPA) and docosahexaenoic acid (DHA), essential FAs, and thus, these FAs need to be provided in the diet. Therefore, fish oils, rich in EPA and DHA, have so far been the main dietary lipid source for Atlantic salmon farming. The supply of available fish oils for farmed fish feed is, however, now limited, and new sustainable alternatives to fish oils are needed in order to encounter this decline. So far, vegetable oils seem to be the most suitable replacement of fish oils in the feed to partially or totally replace fish oil, but the fatty acid profile of plant oils differs from the fatty acid profile of marine oils: fish oils are characterized by high levels of long-chain n-3 polyunsaturated fatty acids (PUFAs) (20 and 22 carbon atoms) and low levels of n-6 fatty acids (FAs), whereas plant oils often contain shorter FAs (18 carbon atoms and less) and high levels of n-6 FAs (Bell et al., 2001; Bell et al., 2002; Olsen et al., 2006; Rosenlund et al., 2001; Thomassen and Røsjø, 1989; Torstensen et al., 2004; Torstensen et al., 2005).

In most studies, changes in muscle fatty acid composition and nutritional quality have been in focus. To what extent substantial changes in dietary fatty acid composition may influence other quality parameters is however not well known. Fatty acid profile of fish flesh changes according to the source of dietary lipids. As long as the essential fatty acids are supplied to the fish, neither growth nor flesh quality seem to be affected (Bell et al., 2001 and 2002; Montero et al., 2005; Nortvedt et al., 2007; Thomassen and Røsjø, 1989; Olsen et al., 2006; Rosenlund et al., 2001), but the integrity of fish metabolism and flesh quality depends on the level of replacement of fish oil by alternative oils and the nature of the fatty acids present in the feed (Montero et al., 2005). Thomassen and Røsjø (1989) found significant effects on odour, taste, colour shade and colour intensity in Atlantic salmon fed a feed with 68 % replacement of fish oil by soybean or rapeseed oil. Izquierdo et al. (2005) showed that at least 60 % replacement of fish oil by soybean oil in seabream diet (diet at 60 % PUFAs among which 50 % were n-3 FAs (% of total FAs)) for 7 months, resulted in a softer flesh. Replacement of fish oil by some alternative lipid sources (vegetable oils (soybean, rapeseed, palm and olive) and animal oil (lard)) influences fat metabolism in rainbow trout (Caballero et al., 2002). In Atlantic cod, the inclusion of soybean oil in the diet caused a faster energy depletion *post-mortem*, faster *rigor* contraction and faster fillet flesh softening, and affected the colour of the flesh (Mørkøre, 2006). It is therefore important to deeply explore the influence of dietary lipids on fish flesh quality and understand the mechanisms behind these changes.

Fatty acid oxidation in fish muscle results in the production of peroxides and aldehydes, possibly giving a rancid taste and smell to the flesh and affecting fish health. Fatty acid oxidation increases exponentially with the number of double bonds of the fatty acids. Lipid oxidation can also occur *in vivo*. The amount of antioxidants in the muscle at slaughter and during processing, and the presence of oxygen and blood residues during processing and storage of the fish can determine the degree of oxidation *post-mortem* (Nortvedt et al., 2007).

1.5.5. Pre-slaughter stress

Pre-slaughter handling stress is an important issue for Atlantic salmon industry because of its effect on welfare and flesh quality. Increase of plasma cortisol levels is the most measured primary physiological response to stress in fish, and increasing levels of cortisol have been associated with higher protein degradation.

During stress, fish use anaerobic energy, resulting in low initial *post-mortem* muscle pH (Bagni et al., 2007; Erikson et al., 1997a; Lowe et al., 1993; Poli et al., 2005; Stien et al., 2005; Thomas et al., 1999); this low initial pH being potentially responsible for the degradation of muscle quality (Lavéty et al., 1988; Nakayama et al., 1996). Direct mechanical stress of the myofibrils or connective tissue, causing the release of proteases, could also contribute to the acceleration of muscle structure degradation (Roth et al., 2006).

If fish are harvested in a rested stage, the energy levels are high and the onset of *rigor* may start as late as 20-30 h after death (Nortvedt et al., 2007). Pre-slaughter stress accelerates the breakdown of ATP, resulting in earlier *rigor* onset and making it difficult to obtain and process *pre-rigor* fillets (Kiessling et al., 2004; Mørkøre et al., 2008; Sigholt et al., 1997; Skjervold et al., 1999). Reducing pre-slaughter stress will increase the potential of *pre-rigor* filleting, thus providing higher freshness and quality products on the market. Pre-slaughter stress has also shown to increase fillet gaping, accelerate fillet softening, affect skin and fillet colour and increase drip loss (Erikson and Misimi, 2008; Kiessling et al., 2004; Morzel et al., 2003; Mørkøre et al., 2008; Roth et al., 2006; Skjervold et al, 2001a; Stien et al., 2005).

Typical transports of fish operate with fish densities of 80-200 kg/m^3, but some transport operators employ even higher densities (Kiessling et al., 2007). Densities can also be high when the fish are crowded on the shore, just before slaughter, and this stressful crowding can seriously affect flesh quality.

1.5.6. Muscle pH

Reported initial muscle pH in unstressed Atlantic salmon immediately *post-mortem* ranged around 7 (Hansen et al., 2007). During further *post-mortem* storage of fish, muscle pH usually decreases in the first stages, because of the degradation of glycogen reserves and consequent production of lactic acid in the muscle, and increases afterwards, due to bacterial production of basic amine (NH_3) (Hansen et al., 2007; Wang and Brown, 1983). *Post-mortem* muscle pH depends on pre-slaughter reserves of glycogen in the muscle. When fish are stressed or subjected to exhausting exercise before slaughter, energy is produced anaerobically, resulting in the production of lactic acid and leading to a low initial *post-*

mortem muscle pH (Bagni et al., 2007; Erikson et al, 1997a; Lowe et al., 1993; Poli et al., 2005; Stien et al., 2005; Thomas et al., 1999). Resting state before slaughter, on the contrary, will spare glycogen reserves, and result in a higher muscle pH immediately *post-mortem*. Low initial muscle pH could be responsible for the deteriorated muscle quality (Lavéty et al., 1988; Nakayama et al., 1996).

1.5.7. Flesh texture and muscle integrity

In Norway, extensive breeding programmes with Atlantic salmon and rainbow trout started in 1971. The breeding goals were focused on growth rate, age of sexual maturation, disease resistance and meat qualities such as fat percentage, fat distribution and flesh colour (Gjedrem, 2000; Johnston et al., 2006; Powell et al., 2008; Quinton et al., 2005), but little attention was given to muscle quality and fillet texture. Selection programmes in several fish species have shown a response of 10-15 % increase in growth rate per generation, and there are indications that similar responses can be obtained for other economically important traits in fish: improve flesh quality will stimulate the consumption of fish by consumers (Gjedrem, 2000).

Fillet texture is one of the most important quality parameters in fish, and in contrast to mammalian meat, consumers generally prefer firm and elastic fish flesh (Rasmussen, 2001). Excessive flesh softness and degradation of the muscle also make the fish fillets too fragile to process (smoking, freezing, slicing, etc.). Today, soft flesh and gaping represent around 40 % of the causes of downgrading losses during the secondary processing of Atlantic salmon (Mitchie, 2001). During the last decade, complaints about gaping and softness during winter-time have increased (Espe et al., 2004), being a serious economical problem for the Atlantic salmon farming industry in Norway and other parts of the world; but no studies have shown conclusive reasons to this soft texture, except a possible influence of harvest season (Johnston et al., 2007).

Fish flesh texture is usually expressed as flesh firmness, measured either instrumentally or by sensory analysis. Atlantic salmon flesh firmness decreases gradually during *post-mortem* storage and can be influenced by several factors: harvesting season, fish species, age and size, length of starvation before slaughter, pre-slaughter handling stress, dietary lipids and feed influencing fat content of the fillet, onset and development of *rigor mortis*, method of slaughter, fillet processing and storage temperature (Bagni et al., 2007; Einen and Thomassen, 1998; Einen et al., 2002b; Guillerm-Regost et al., 2006; Izquierdo et al., 2005; Kiessling et al., 2004; Mørkøre and Rørvik, 2001; Mørkøre et al., 2008; Skjervold et al., 2001a and 2001c). Factors including amount and properties of the proteins, muscle collagen content and disintegration, muscle cellularity and rate and extent of proteolysis causing breakdown of myofibrils and connective tissue can also be determining (Ando et al., 1991, 1992a and 1992b; Godiksen et al., 2009; Hultmann and Rustad, 2004; Johnston, 1999; Johnston et al., 2000 and 2006; Taylor et al., 1995 and 2002; Yamashita and Konagaya, 1990). Texture also appears closely related to the pH of the flesh (Rasmussen, 2001).

Muscle integrity and degradation determine fish flesh quality. Muscle histology allows a direct observation of myofibre and connective tissue degradation. During *post-mortem* storage, fish muscle degrades gradually. Ando et al. (1991) and Ofstad et al. (2006) suggested the gradual degradation of the extra-cellular matrix may be one of the causes of *post-mortem* tenderization of fish muscle. Taylor et al. (2002) have also shown that breaks in muscle cell cytoskeleton and connective tissue (detachments between myofibres and between myofibres and myocommata) determined fish flesh texture. They suggested that fish muscle tenderization paralleled loss of attachments between myofibres.

The action of proteases like lysosomal cathepsins and/or cytosolic calpains can be responsible for the degradation of myofibres and connective tissue, and protease activity has been associated with fish muscle softening (Aoki et al., 1997; Yamashita and Konagaya, 1990 and 1991). Collagen content and its disintegration, as well as the concentration of immature and mature cross-link species that link collagen molecules in the connective tissue matrix can also contribute to fish flesh texture (Ando et al., 1991, 1992a and 1992b; Espe et al., 2004; Johnston, 1999; Johnston et al., 2006). Insoluble collagen can be attacked by collagenases, neutral proteinases (calpains) and/or acid proteinases (cathepsins), cleaving the triple helix (Ando et al., 1992a).

2. Materials and Methods

Atlantic salmon farming is a relatively new but fast growing industry, which has to face challenges: to be able to provide the consumer with a high quality product is one. Atlantic salmon flesh quality can be influenced by external and internal factors, both *pre-mortem* and *post-mortem*. The aim of the three studies (I: Bahuaud et al., 2008; II: Bahuaud et al., 2009b; and III: Bahuaud et al., 2009a) was to investigate the impact of three main factors on Atlantic salmon flesh quality; these factors being directly related to the actual challenges encountered by the industry. The potential role of lysosomes and lysosomal enzymes, cathepsins, in maintaining or degrading Atlantic salmon muscle integrity in relation with these factors, was in focus. In more details, we aimed at:

- Studying the impact of super-chilling on Atlantic salmon flesh quality, including muscle structure integrity, fillet texture, liquid leakage, as well as the release and activity of lysosomal cathepsins. In this context, we aimed at emphasizing the role of these cathepsins, in the *post-mortem* degradation of salmon flesh, after super-chilling of the fillets (Study I: Bahuaud et al., 2008).
- Investigating the effect of dietary rapeseed oil, rich in n-6 fatty acids, on Atlantic salmon flesh quality measured as muscle structure and lysosomal cathepsin total activity. The role of lysosomes and cathepsins, in relation to lipid metabolism and flesh quality, was studied with the incorporation of very high levels of n-3 fatty acids (EPA and DHA) in Atlantic salmon diets (Study II: Bahuaud et al., 2009b).
  Evaluating the effect of pre-slaughter short- and long-term crowding stress on Atlantic salmon muscle quality, measured as muscle structure integrity and fillet texture; and to study the potential role of muscle pH and lysosomal enzymes, cathepsins, in *post-mortem* degradation of fish muscle, in relation to pre-slaughter stress (Study III: Bahuaud et al., 2009a).

Three different and independent studies (I: Bahuaud et al., 2008; II: Bahuaud et al., 2009b; and III: Bahuaud et al., 2009a) are thus reported in this chapter. Some methods of sample analyzing were common to all three studies and are described in this section. The results of the three studies were however obtained and analyzed separately.

2.1. Fish and Experimental Settings

2.1.1. Study I: Super-chilling of atlantic salmon fillets (Bahuaud et al., 2008)

Twenty-four Atlantic salmon (*Salmo salar*) were raised at Nofima Marin (previously Akvaforsk) Research Station, Averøy, on the west coast of Norway. They were fed a standard commercial extruded dry feed ("Optiline", 12mm, Skretting, Stavanger, Norway). The salmon were starved for five days before sampling (sea temperature 8 °C, salinity 33 $^0/_{00}$, salmon average body weight of 5 kg). After killing by percussive stunning, gill cutting and subsequent bleeding in a tank with running water, the salmon were filleted immediately *post-mortem* by experienced workers (*pre-rigor* fillets) and subjected to two different treatments. Each right salmon fillet was placed in a mobile super-chilling tunnel, in a freezing room at a temperature of -25 °C, with an air velocity of approximately 2.5 m/s in the tunnel, above the fillets. The fillets were super-chilled for 45 minutes, after which the temperature of the fillet cores had reached -1.5 °C. The super-chilled fillets were immediately packed on ice in standard Styrofoam boxes and stored in a cold room (+5 °C) for further analyses, until 144 h after slaughter. Each left salmon fillet served as control and was directly packed on ice, the same way as super-chilled fillets, and placed in the cold room, but without previous super-chilling treatment.

Table 1. Formulation and chemical composition of the diets (Bahuaud et al., 2009b).

Formulation (% of total weight)	FO	RO	EPA	DHA
Fish meal, LT	67.9	67.9	67.9	67.9
Fish oil	13.5			
Rapeseed oil		13.5		
EPA[1]			13.5	
DHA[2]				13.5
Wheat	17.1	17.1	17.1	17.1
Vitamin premix[3]	1.0	1.0	1.0	1.0
Mineral premix[c]	0.4	0.4	0.4	0.4
Yttrium oxide[4]	0.01	0.01	0.01	0.01
Carophyll Pink[5] (8%)	0.064	0.064	0.064	0.064
Chemical composition				
Dry matter (%)	92.3	92.2	93.0	92.2
% of dry matter				
Crude proteins	55.4	54.6	55.1	54.4
Crude fat	22.5	23.0	22.7	23.3
Ash	9.5	9.9	9.5	9.7
Gross energy (MJ/kg)	23.8	23.9	23.6	23.8
150 ppm BHT were added to the oils used in the diets. LT=Low temperature				

[1] Incromega EPA 500TG SR, Croda Chemicals Europe Ltd., Goole, England
[2] Incromega DHA 500TG SR, Croda Chemicals Europe Ltd., Goole, England
[3] Mundheim et al. (2004)
[4] Inert marker, Y2O3, Sigma
[5] Hoffman-LaRoche, Basel, Switzerland

Besides the ice-stored fillets (super-chilled and control), pieces of *pre-rigor* fillets (15 x 10 cm) from 5 other salmon, were packed in modified atmosphere (MAP) and stored in the cold room (+5 °C) for long term analyses, until 672 h (4 weeks) after treatment. Before being MAP packed, five of these right fillets were also previously super-chilled. The other five left fillets were immediately MAP packed without previous super-chilling and served as control for the rest of the experiment with MAP. In this study, MAP was mainly used to extend shelf-life of the fillets and be able to run long term analyses with law bacterial disturbance, and the effect of MAP was not reported in this chapter.

Super-chilling formed an ice layer on the top of the fillets of approximately 4 mm. After previous testing, it was estimated that about 25 % of the free water contained in the fish was frozen at the end of the super-chilling treatment.

2.1.2. Study II: Dietary n-3 and n-6 fatty acids (Bahuaud et al., 2009b)

Four groups of Atlantic Salmon (*Salmo salar*) with an initial average weight of 90 g, were distributed in 3 cylinder-conical (0.85 m diameter) fresh water tanks per group (temperature of the water 10.1 °C) at Nofima Marin Research Station (previously Akvaforsk), Sunndalsøra, Norway. Four different extruded diets (Table 1) were prepared with 50 % (of total weight) crude proteins and 23 % (of total weight) crude lipids, among it 100 % of the added oils were either fish oil (FO), rapeseed oil (RO), eicosapentaenoic acid (EPA)-enriched-oil or docosahexaenoic acid (DHA)-enriched-oil.

Each salmon group was fed one of the four diets for 21 weeks until reaching an average weight of 344 g (± 69.8). Three fish per group taken randomly from the different tanks were anaesthetized in metacain (MS-222) provided by Norsk Medisinaldepot (Norway), killed with a blow to the head, gill cut and stored in a refrigerated room (+4 °C) for 96 h until further analyses.

2.1.3. Study III: Short-term and long-term stress (Bahuaud et al., 2009a)

Atlantic salmon (*Salmo salar L.*) were raised in sea-water at Nofima Marin (previously Akvaforsk) Research Station, Averøy, on the west coast of Norway, where they were fed a standard commercial extruded dry feed ("Optiline S 800", 9 mm, Skretting, Stavanger, Norway). Three weeks before slaughter, 600 salmon were randomly distributed into 4 net-pens (125 m^3) in the sea, each of them containing 150 fish. The salmon (average body weight of 2.9 kg ± 0.1) were starved for 4 days before slaughter (sea temperature 12.7 °C, salinity 33 $^0/_{00}$) and were subjected to 3 different pre-slaughter handling procedures: Minimal crowding Stress (NS) (2 net-pens), Short-term crowding Stress (SS) (1 net-pen), or Long-term crowding Stress (LS) (1 net-pen).

Salmon from the LS group were kept crowded in their net-pen by reducing the size of the net-pen in order to obtain a fish density of 300 kg/m^3, for about 24 h from the day before slaughter. At the day of slaughter, salmon from each group were transferred from their net-pen to a tank with refrigerated sea-water (RSW 2.0 °C ± 0.1), where they were live-chilled and kept at either a low density of 65 kg/m^3 (NS group) or high density of 300 kg/m^3 (SS and LS groups) for 20 min. Salmon were then killed by percussive stunning, gill cut and bled in a tank with running RSW for 20 min. All the salmon used for analyses were collected in 2 harvests per group. To avoid any additional disturbance, 2 net-pens were attributed to the NS group only, so that each net-pen in this group had to be harvested only once.

Table 2. Summary of analyses and measurements performed on salmon muscle in the three studies (I, II and III).

Analyses and measurements	Study I: Super-chilling (Bahuaud et al., 2008)	Study II: Dietary lipids (Bahuaud et al., 2009b)	Study III: Pre-slaughter stress (Bahuaud et al., 2009a)
Treatment groups compared	Super-chilled vs. Control fillets	Muscle from 4 diets: FO, RO, EPA and DHA	Fillets from salmon subjected to 3 lengths of pre-slaughter crowding stress: NS, SS and LS
Number of salmon muscle or fillets analysed / treatment group	6	3	6
Myofibre-myofibre detachments during storage	X	X	X
Myofibre-myocommata detachments during storage	X	X	X
Myofibre breakages at 144h post-mortem	X	X	X
Contracted myofibres at 144h post-mortem	X	X	X
Cathepsin B and cathepsin L activity in the muscle lysosomal fraction during storage	X		
Cathepsin B and cathepsin L total activity in the muscle homogenate at 0h post-mortem		X	X
Cathepsin B and cathepsin L muscle relative gene expression at 0h post-mortem		X	X
Instrumental fillet texture (firmness) during storage time	X		X
Liquid leakage at 168h post-mortem	X		
Lipid class and fatty acid analyses in the muscle CLF (Crude Lysosomal Fraction) at 0h post-mortem		X	
Muscle pH during storage time			X
Statistics	X	X	X

Seventy-two salmon were taken from each net-pen and killed for the present study; the rest of the salmon in the different net-pens being used for other studies not mentioned in this chapter. For the present study, the killed salmon (n=72) were distributed in 4 batches (n=18 in each batch, 6 salmon in each group NS, SS and LS). The first 3 batches were hand-filleted immediately *post-mortem* by experienced workers to obtain "*pre-rigor* fillets" that were placed on solid plastic trays in closed Styrofoam boxes with ice, and stored in a cold room (5

°C) for further analyses. Salmon from the last batch were packed as gutted fish on ice in Styrofoam boxes and stored in a cold room (+5 °C) until they were filleted *post-rigor*, 96 h after slaughter, and analysed as "*post-rigor* fillets". Only the left fillets were used for sampling and analyses in the present study. Each left fillet from each batch of salmon was used for sampling and analyses on maximum 3 locations at 3 different time-points *post-mortem*.

2.2. Analyses of Quality and Atlantic Salmon Muscle Structure

Table 2 gives a summary of all the analyses and measurements performed and reported in this chapter, in the three studies (I, II and III). The methods used for quality and muscle structure analyses and measurements are described in more details by Bahuaud et al. (2008, 2009b and 2009a).

3. Results and Discussion

3.1. Measurement of Cathepsin Activity in the Different Muscle Fractions

Cathepsin activity can be measured in different fractions of the muscle. In studies II and III, cathepsin activity was measured only as total, in the muscle homogenate. Cathepsin activity has been measured as total activity in muscle homogenates in several other studies (Dunn and Rustad, 2008; Chéret et al., 2006 and 2007; Mommsem, 2004; Salem et al., 2006). Measuring cathepsin total activity in muscle homogenate immediately *post-mortem* highlights the level of potential muscle proteolysis existing already *in vivo* before death and following during storage time. Cathepsin activity was measured as total activity in the muscle homogenate when the factor potentially influencing *post-mortem* quality was applied when the fish were still alive. We evaluated the effect of different dietary lipids in the feed (study II, Bahuaud et al., 2009b) and pre-slaughter stress (study III, Bahuaud et al., 2009a) on flesh quality measured as muscle cathepsin total activity, these factors applied before death, potentially influencing the level of enzyme activity already when the fish were still alive. A higher cathepsin total activity in the muscle homogenates thus indicate higher potential muscle degradation *in vivo* before death and consequently during storage time *post-mortem*.

In study I, cathepsin activity was measured in the muscle lysosomal fraction (pellet). Measuring cathepsin activity in the lysosomal fraction of the muscle indicates the degree of release of cathepsins from the lysosomes to the cytosol during storage time (Aoki and Ueno, 1997) and thus, the degree of potential following muscle degradation. The release of lysosomal enzymes, cathepsins, to the cytosol, follows the breakage of the lysosomal membrane (Allison, 1975; Béchet et al., 2005; Kölher, 1991; Moore, 1988; Moore and Lowe, 1985). Cathepsin activity was measured in the lysosomal fraction of the muscle when the factor potentially influencing *post-mortem* quality was applied after death. We thus evaluated the effect of super-chilling of Atlantic salmon fillet (study I, Bahuaud et al., 2008) on quality, measured as cathepsin activity in the muscle lysosomal fraction (pellet). A decrease of cathepsin activity in the muscle lysosomal fraction (pellet) during storage time indicates a progressive breakage of the lysosomal membrane, leading to the release of these lysosomal proteolytic enzymes, and eventually the start of muscle degradation.

3.2. Study I: Super-Chilling (Bahuaud Et Al., 2008)

3.2.1. Ice crystal formation

Light microscopy observations were performed in order to detect eventual physical damages to the muscle, due to super-chilling. Observations of the upper layer of the fillets showed several freezing damages in all super-chilled fillets, apparently due to the development of large intra- and extra-cellular ice crystals during super-chilling (Figure 8). Intra-cellular ice crystals seemed to be formed mostly at the border of the myofibres, while extra-cellular one's affected the endomysium. In the control fillets, no intra- or extra-cellular ice crystals formation traces were noticed.

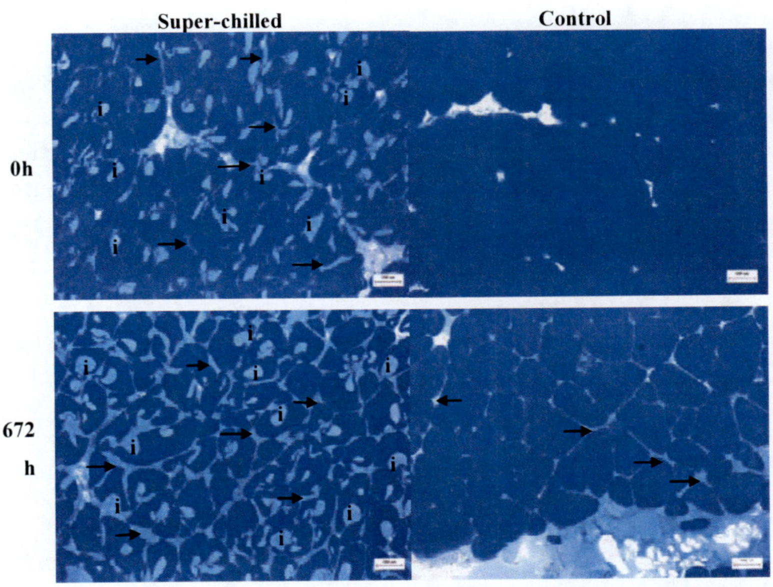

Figure 8. Microscopic observations (x10) of Atlantic salmon muscle at 0 h and 672 h (MAP) *post-mortem*. Arrow = myofibre-myofibre detachments; i = ice crystals. Scale bar = 100 µm (Bahuaud et al., 2008).

Molina-Garcia et al. (2004) found that classical freezing has as consequences, disruption and separation of muscle fibre bundles in pork meat due to the occurrence of extra-cellular and numerous intra-cellular ice crystals. On the other hand, they found that super-cooled meat did not show any freezing damages (the samples lacked intra-cellular ice crystals and the muscle fibre bundles were undisturbed) and even preserved the quality of the muscle. They explained these results by the difference in freezing volume between classical freezing and super-cooling, classical freezing inducing bigger ice crystals than super-cooling. In our study, the ice crystals formed during super-chilling were large size ice crystals and big enough to damage the integrity of the salmon muscle. It thus seems that -1.5 °C short super-chilling of

Atlantic salmon fillets resulted in similar damages as the ones observed in both -20 °C classical freezing and subzero super-chilling of fish and/or meat. Extra-cellular ice crystal formation was also previously noticed by Choi and Park (1991) in -3 °C super-chilled yellowtail fish. Lee and Toledo (1984) however found that ice crystals formed in subzero temperature stored (-2 °C) mullet caused less structural changes than in -20 °C frozen mullet, which presented destructive ice crystals in the muscle.

The optimum freezing rate for a food product thus depends a lot on the freezing system. Slow classical freezing usually forms extra-cellular ice crystals, decreasing the product quality by extensive mechanical damages, accelerated enzyme activity and increased oxidation rates, whereas ultra-rapid freezing can reduce ice crystal size but still may result in intra-cellular ice crystallisation and/or mechanical cracking in the tissues. Intra-cellular ice crystals can affect the internal structure of the muscle (Molina-Garcia et al., 2004) and cause protein denaturation (Gomez and Calvelo, 1982; Martino and Zaritzky, 1988). The size and location of the ice crystals have shown to affect the textural quality of a frozen food, as well as drip loss after thawing (Bello et al., 1982; Martino et al., 1998).

3.2.2. Myofibre-myofibre Detachments

Figure 9 shows the percentages of myofibre-myofibre detachments in Super-chilled and Control salmon fillets during storage time. In the cross-section muscle samples, an increase in myofibre-myofibre detachments with storage time was noticed, up to 144 h after treatment for both super-chilled and control fillets ($p<0.0001$), and this illustrates the normal fish mucle degradation process during storage (Ofstad et al., 2006; Taylor et al., 2002).

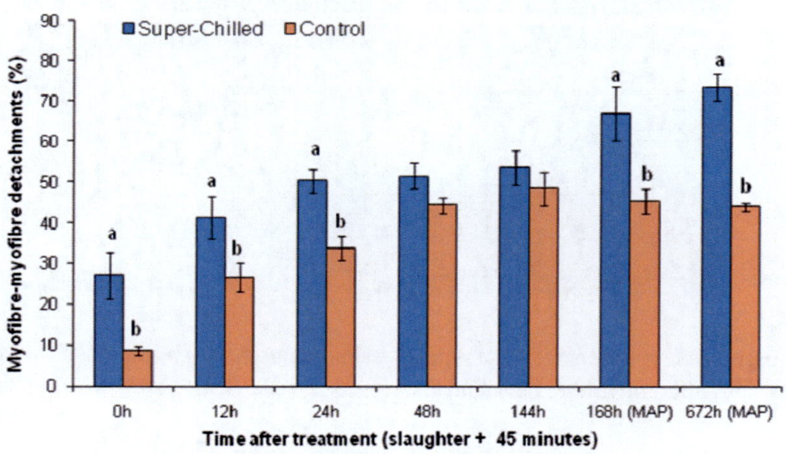

Figure 9. Average percentages of myofibre-myofibre detachments in Super-Chilled and Control Atlantic salmon *pre-rigor* fillets during storage time. Each histobar represents the mean value ±SE of one treatment group (n=6 in each group) at one time-point after treatment. Different letters denote significant differences between treatments at the same time-point ($p<0.05$).

Super-chilling caused more myofibre-myofibre detachments than control during storage time ($p<0.0001$). Already immediately after treatment (0 h samples), super-chilled fillets presented significantly more myofibre-myofibre detachments than the control fillets ($p=0.009$), and the same phenomenon could still be seen at the end of the storage period (672 h after treatment) ($p<0.0001$). The acceleration of myofibre-myofibre detachments in our super-chilled fillets seemed to be a direct consequence of freezing damages, confirmed by the large intra- and extra-cellular ice crystals formed during super-chilling.

3.2.2. Myofibre-myocommata detachments

Figure 10 shows the percentages of myofibre-myocommata detachments in Super-chilled and Control salmon fillets during storage time. In contrast to myofibre-myofibre detachments, no significant difference or clear trend was noticed in myofibre-myocommata detachment level between treatment (super-chilling and control) or storage time (from 0 h to 672 h after treatment). These results agreed with Ofstad et al. (2006) and Taylor et al. (2002), who did not notice any changes in myofibre-myocommata attachments before 5 days of storage in Atlantic salmon and 7 days in wolfish and confirm that myofibre-myocommata detachments apparently occur later than myofibre-myofibre detachments in these fish species.

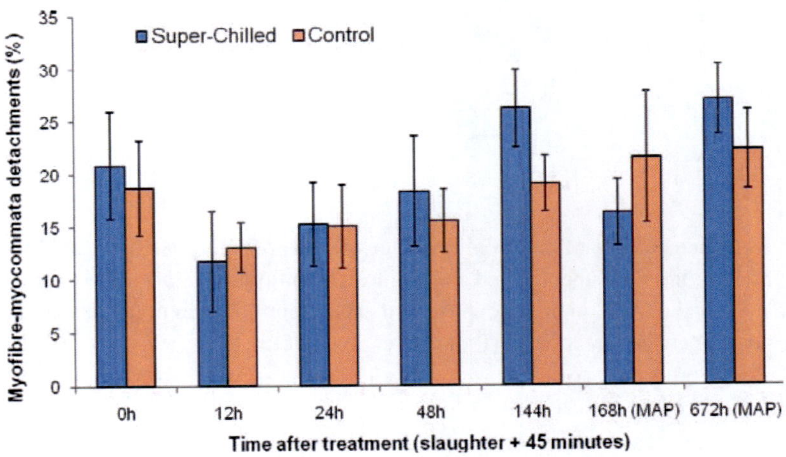

Figure 10. Average percentages of myofibre-myocommata detachments in Super-Chilled and Control Atlantic salmon *pre-rigor* fillets during storage time. Each histobar represents the mean value ± SE of one group at each time-point after treatment (n=6 in each group).

Figure 11. Microscopic observations (x10) of Atlantic salmon muscle at 144 h after treatment. b = myofibre breakages; i = ice crystals; c = contracted myofibres. Scale bar = 100 µm (Bahuaud et al., 2008).

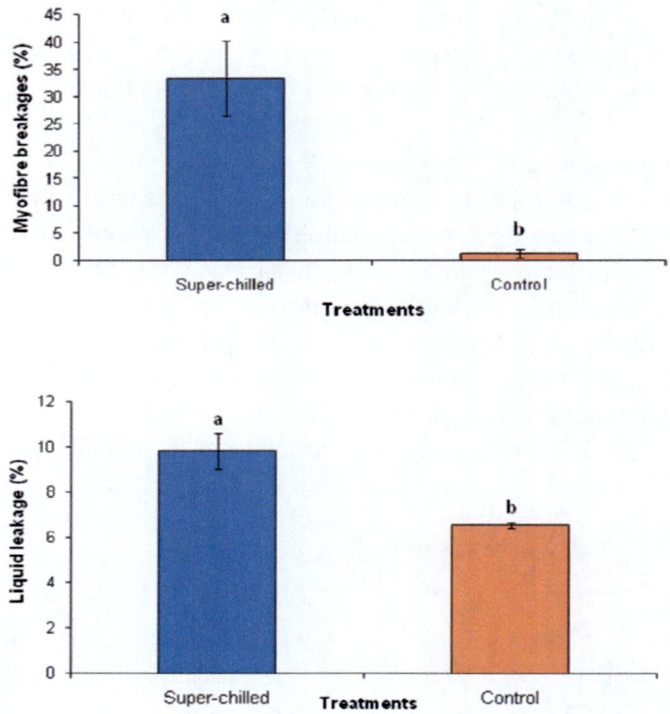

Figure 12. Average percentages of myofibre breakages 144 h *post-mortem* and liquid leakage 168 h *post-mortem* (MAP fillets) in Super-chilled and Control Atlantic salmon *pre-rigor* fillets. Each histobar represents the mean value ± SE in one group (n=6 in each group). Different letters denote significant differences ($p<0.05$) between treatments (Bahuaud et al., 2008).

3.2.4. Myofibre breakages and liquid leakage

Microscopic observations of longitudinal sections of the super-chilled fillets 144 h after treatment showed a significant higher percentage ($p=0.005$) of myofibre breakages comparing to the control fillets (Figures 11 and 12), which agreed with our results on myofibre-myofibre detachments whose level was accelerated by super-chilling. According to Sanz et al. (1999), formation of ice crystals in muscle dehydrate the fibres, resulting in a higher concentration of salt in the tissue. This dehydration could lead to denaturation of the muscle proteins and structural damages of the membranes (Huss, 1988).

Liquid leakage 168 h after treatment (MAP fillets) was significantly higher in the super-chilled fillets than in the control fillets. Sivertsvik et al. (2003) observed less drip loss in -2 °C super-chilled stored MAP fillets than in control MAP fillets, but in their study, no ice crystals were formed in the upper layer of the super-chilled fillets. Our results agreed more with the ones from Einen et al. (2002a), who found a higher drip loss in -25 °C frozen Atlantic salmon *pre-rigor* fillets, compared to fresh fillets; classical freezing in this case probably forming ice crystals. Gallart-Jornet et al. (2007) also found a decrease in water holding capacity of -1 °C super-chilled Atlantic salmon *post-rigor* fillets and explained it by possible structural changes in the muscle during super-chilled storage. Einen et al. (2002a), Mackie (1993) and Sikorski and Kolakowska (1994) also suggested that a lower water binding capacity in fish muscle can be due to cell damages, lower protein solubility and protein denaturation and aggregation during freezing and thawing. These explanations could link our results on liquid leakage and myofibre breakages together, suggesting that myofibre breakages due to formation of ice crystals in the muscle could be responsible for the higher liquid leakage in super-chilled fillets.

Table 3. Average percentage ±SE of contracted myofibres and breaking force (firmness) in Super-chilled and Control Atlantic salmon *pre-rigor* fillets, 168 h after treatment (MAP). Different letters denote significant differences between treatments ($p<0.05$).

	Contracted myofibres (%)	Breaking force (Fillet firmness) (N)
Super-chilled	27.8^{b} ±7.3	8.98 ±1.02
Control	53.3^{a} ±12.6	8.95 ±0.69

3.2.5. Contracted myofibres and fillet firmness

Contracted muscle fibres observed microscopically on longitudinal muscle sections were found in almost twice as often seen in the control fillets as in the super-chilled fillets ($p=0.04$) (Table 3). In Atlantic salmon, Mørkøre et al. (2007) measured only 5 % of *rigor* contraction in super-chilled *pre-rigor* fillets and 18 % in control fillets. Accordingly, Einen et al. (2002a) showed that nitrogen frozen *pre-rigor* Atlantic salmon fillets did not develop any thaw *rigor* muscle contraction whereas unfrozen *pre-rigor* fillets contracted about 14 % in length during *rigor*; and this absence of contraction had a negative effect on the final quality of the fillet, such as decrease in the firmness of the fillets.

In our study, there was however no difference in fillet firmness measured instrumentally as breaking force, between super-chilled and control fillets 168 h after treatment (MAP fillets) (Table 3). No correlation over time between firmness of the fillets (results not shown) and level of myofibre-myofibre detachments on one hand, and level of cathepsin B+L activity in the lysosomes on the other hand, was observed either. Texture is determined by several distinct structures and events, among them levels of myofibre-myofibre and myofibre-myocommata attachments/detachments (Taylor et al., 2002). We found more myofibre-myofibre detachments in the super-chilled group at this time point, a higher release of cathepsin B+L from the lysosomes (lower cathepsin B+L activity in the lysosomal fraction) 168 h after treatment, lower myofibre contraction, as well as more myofibre breakages 144 h after treatment in the super-chilled fillets. According to these results, we expected softer super-chilled fillets 168 h after treatment. On the other hand, we observed the same percentage of myofibre-myocommata detachments in both groups and a large variation between fillets texture results (breaking force), especially in the super-chilled group. Besides, at this time point, both fillet groups presented more than 45 % of myofibre-myofibre detachments. At this level of deterioration, it is possible that a threshold has been reached, above which no differences in texture could be measured instrumentally. It was however interesting to notice that at this same time-point, two of the super-chilled fillets showed a very soft texture reaching the limit of acceptation (5.31 and 6.93 N) for fish quality (Mørkøre, 2008), but no extreme results like these were detected in the control fillets.

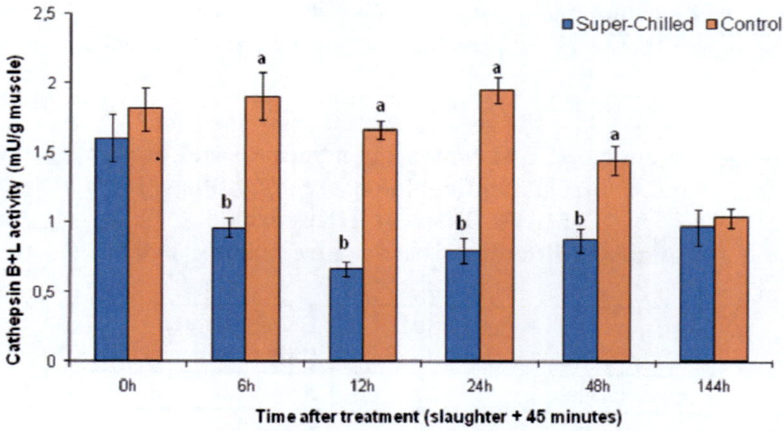

Figure 13. Average cathepsin B+L activity in the lysosomal fraction (pellet) of super-chilled and control Atlantic salmon *pre-rigor* fillets. Each histobar represents the mean ±SE in each group at each time-point after treatment. Different letters denote significant differences (p<0.05) between treatments (n=6 in each group) at one time-point.

3.2.6. Cathepsin B+L activity in the lysosomal fraction

Cathepsin B+L activity significantly decreased during storage time in the lysosomal fraction (p<0.0001) of both super-chilled and control fillets (Figure 13), which was also observed by Aoki and Ueno (1997) in mackerel muscle. This decrease corresponds to the normal degradation process of the muscle during storage, cathepsins being released from the lysosomes to start proteolysis.

Super-chilling caused a higher release of cathepsin B+L from the lysosomes (p<0.000001). Already 6 h after treatment, almost half of the activity had disappeared from the lysosomes of the super-chilled fillets, when compared to 0 h after treatment. On the contrary, the activity in the lysosomes of the control fillets did not start to decrease until between 24 h and 48 h after treatment. A significant negative correlation was found between the level of myofibre-myofibre detachments and the level of cathepsin B+L in the lysosomal fraction (-0.711). The super-chilled fillets also showed a much lower cathepsin B+L activity than the control fillets 168 h after treatment (MAP fillets). The same significant difference could still be observed 672 h after treatment (MAP fillets), although the remaining activity was very low in both groups (results not shown).

Our results on cathepsin B+L activity in the lysosomal fraction are not in line with the results of Gallart-Jornet et al. (2007), whose analysis indicated that -1 °C super-chilling and storage of Atlantic salmon for 9 and 16 days did not cause any lysosomal membrane damages. These damages were evaluated in the centrifugal cell tissue fluid, with the level of lysosomal alpha-glucosidase activity, efficient in detection of freeze-thaw treatment of fish. In our study, we instead chose to measure cathepsin B+L activity in the lysosomal fraction, more appropriated to evaluate muscle softening, as cathepsins have shown to be directly responsible for fish muscle degradation. Accordingly to Gallart-Jornet et al. (2007), Losada et al. (2005) observed that -1.5 °C super-chilling storage of gutted *post-rigor* horse mackerel using slurry ice reduced nucleotide autolytic degradation during storage time. But the super-chilling technique they used did not imply any freezing damages and formation of ice crystals, which would explain the absence or reduction of degradation in their super-chilled fish. Duun and Rustad (2008) also concluded that super-chilling did not influence cathepsin B+L activity in the muscle (at -1.4 °C and -3.6 °C); this activity being measured as total activity in the muscle homogenate. To our opinion, measuring cathepsin B+L total activity would not give any indication about the impact of super-chilling on fillet quality, as the total enzyme activity should not vary from one treatment to the other if the treatment is done *post-mortem*.

Our results suggested that super-chilling affected the integrity of the muscle lysosomes and consequent release of cathepsins, more rapidly than the control treatment. This agreed with our results of myofibre-myofibre detachments, which showed that super-chilling accelerated muscle deterioration. Muscle autolysis of mature chum salmon due to the activity of cathepsin B and L has shown to be enhanced if the muscle proteins were previously denaturised (Yamashita and Konagaya, 1991). The ice crystals damaging myofibres in the upper layer of our super-chilled fillets could thus be one of the causes of the accelerating release of proteolytic enzymes in the super-chilled group. Our results on cathepsin activity in the lysosomal fraction could also be correlated with those found on yellowtail fish by Choi and Park (1991), who showed that -3 °C partial freezing resulted in deleterious changes of the mitochondrial inner membrane.

3.3. Study II: Dietary Lipid (N-6 and N-3 Fatty Acids) (Bahuaud Et Al., 2009b)

3.3.1. Fish and diets

No significant differences in weight (p=0.683) and length (p=0.269) were observed between the dietary groups at the day of sampling.

An overview of the fatty acid composition of the four diets is described in Figure 14. The FO and RO diets were characterized by the highest levels of saturated fatty acids (SFAs) and monounsaturated fatty acids (MUFAs) respectively, whereas the EPA and DHA diets contained almost twice their level of PUFAs, reaching more than 60 % of the total fatty acids, with 50 % consisting of EPA+DHA. By feeding Atlantic salmon with the RO diet, we aimed at evaluating the impact of this diet on muscle structure stability. Feeding salmon with the EPA and DHA diets, considered as extreme diets potentially representing a stress for the fish, allowed us to highlight the potential implication of lysosomes and lysosomal enzymes, cathepsins, in salmon muscle degradation, in relation to dietary fatty acids. The FO diet served as control for comparison of the effects with the other diets.

3.3.2. Lipid class and total fatty acid compositions of the muscle Crude Lysosomal Fraction (CLF)

Figure 15 gives an overview of the total fatty acid composition (% of total fatty acids) of the muscle CLF of salmon fed the four diets (FO, RO, EPA and DHA).

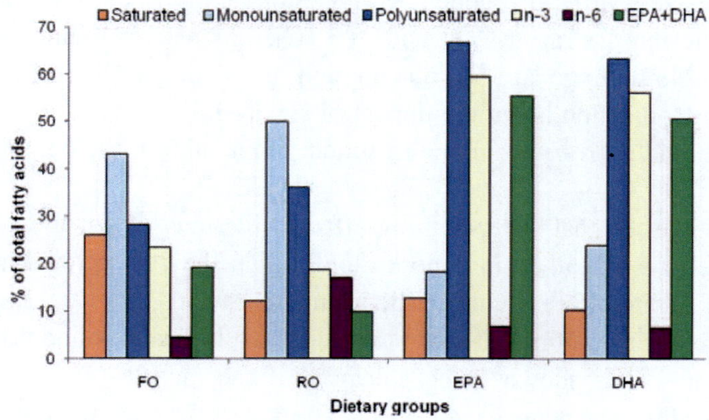

Figure 14. Overview of fatty acid composition of the diets (% of total fatty acids).

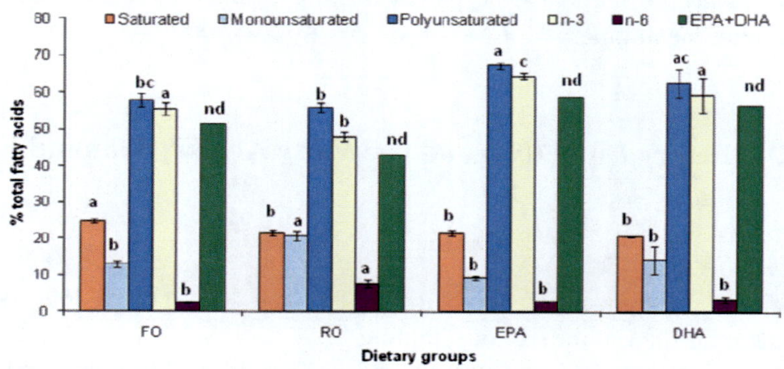

Figure 15. Overview of fatty acid composition of Atlantic salmon muscle lysosomal fraction in the different dietary groups, 0 h *post-mortem*. Each histobar represents the mean ±SE for one dietary group. Different letters denote significant differences ($p<0.05$) between dietary groups (n=3 in each group). nd = p values not determined.

In general, fatty acid composition of the muscle CLF reflected the composition of the diets in all groups. This was not surprising, since we (Thomassen and Røsjø, 1989) and several other groups (Rosenlund et al., 2001; Einen et al., 2002b; Bell et al., 2003; Torstensen et al., 2004; Torstensen et al., 2005; Nanton et al., 2007; Pratoomyot et al., 2008) have shown that fatty acid composition of Atlantic salmon muscle lipids is closely influenced by dietary fatty acids. To the best of our knowledge, however, this study was the first one describing the influence of dietary lipids on fatty acid composition of the muscle CLF (Bahuaud et al., 2009b).

The differences in fatty acid composition observed in the muscle CLF between the FO and RO groups were not as high as found between the FO and RO diets. Replacing fish oil (FO) by rapeseed oil (RO) in salmon diet reduced the percentage of SFAs and increased the percentage of MUFAs in the muscle CLF. The percentage of total n-6 FAs was also increased, but the percentage of n-3 FAs and the ratio n-3/n-6 decreased. The percentage of EPA in the muscle CLF of the RO group was however not significantly different from the FO group, and the percentage of DHA was only 6 % lower. This is most probably related to the CLF fraction being dominated by phospholipids (about 65 %) (data not shown). It is generally accepted that very long chain n-3 FAs, and especially DHA, are preferentially incorporated in phospholipids and that phospholipids are less influenced by dietary fatty acid composition as compared to triacylglycerols (Brodtkorb et al., 1997; Olsen and Henderson 1997).

Replacing dietary FO by an EPA- or DHA-enriched oil also decreased the percentage of SFAs but increased the percentage of PUFAs in the muscle CLF of salmon belonging to these groups, especially the n-3 FAs. These effects were more pronounced with the EPA than with the DHA diet.

No significant differences were noticed in percentage of muscle CLF total phospholipids or neutral lipids between the four dietary groups. Some significant differences were however observed for some individual phospholipids (data not shown). No cardiolipin (CL) was detected in the muscle CLF of salmon fed the EPA or DHA diet, whereas 2.6 and 2.4 % of CL were found in the FO and RO groups, respectively. Cardiolipin is an important component of the inner mitochondrial membrane. The absence of cardiolipin in the muscle CLF of salmon fed the EPA or DHA diet may indicate an increased incidence of lipid peroxidation and oxidative stress, and disrupted mitochondrial membranes (Petrosillo et al., 2001; Kirkland et al., 2002). In the same study using the same salmon as in our study, Kjær et al. (2008) and Todorcevic et al. (2008) observed evidence of excessive oxidative stress with less major phospholipids in mitochondrial membranes, higher membrane damages and a lack of mitochondrial β-oxidation capacity in liver and adipose tissue of salmon fed the EPA or DHA diet. Kurz et al. (2008) suggested a close relationship between lysosomes and mitochondria, where oxidative stress due to production of hydrogen peroxide by mitochondria will impair lysosomes; eventually releasing lysosomal enzymes that will again damage the mitochondria and induce more oxidative stress. Consequently, damages observed in mitochondria may also be found at the lysosomal level; suggesting that the EPA and DHA diets may have caused impairment of the lysosomal function as well.

3.3.3. Microscopy of the muscle

a Myofibre-myofibre detachments

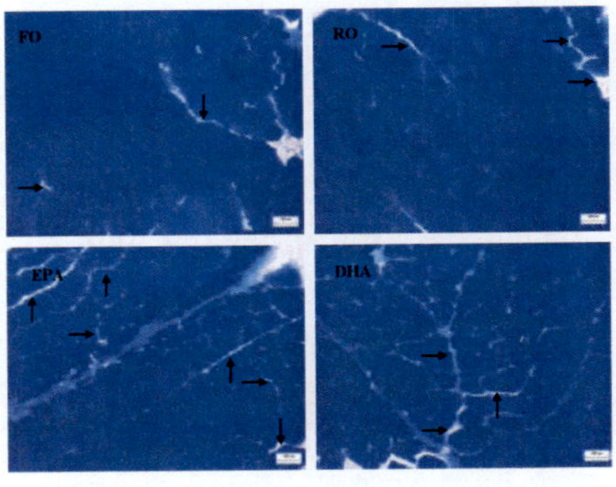

Figure 16. Microscopic observations of Atlantic salmon muscle (x20) at 12 h *post-mortem*, in the different dietary groups (FO, RO, EPA and DHA). Arrows: myofibre-myofibre detachments. Scale bar = 100 µm (Bahuaud et al., 2009b).

Pictures from light microscopy observations of salmon muscle at 0 h *post-mortem* are presented in Figure 16. Figure 17 shows the percentage of myofibre-myofibre detachments during storage time. In general, myofibre-myofibre detachments increased with time ($p<0.0001$) in the four dietary groups until 24 h *post-mortem*. However, no differences in level of myofibre-myofibre detachments were noticed between 24 h and 96 h *post-mortem*.

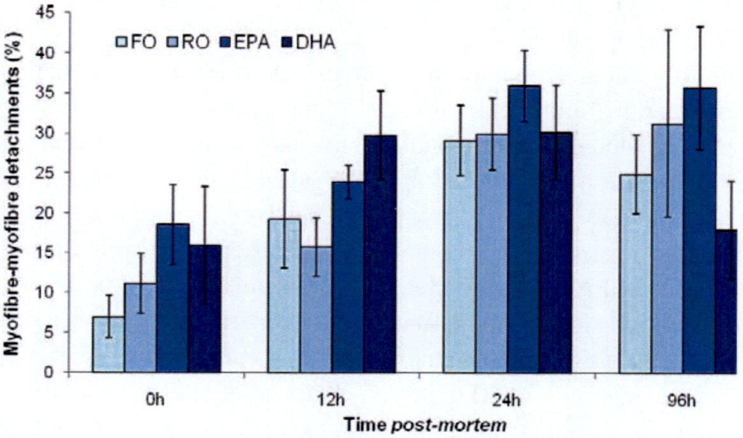

Figure 17 Average percentages of myofibre-myofibre detachments in muscle of Atlantic salmon fed the different diets (FO, RO, EPA and DHA), during storage time. Each histobar represents the mean value ±SE of one dietary group (n=3 in each group) at each time-point (Bahuaud et al., 2009b).

Overall, each time-point taken individually, it seems that the FO and RO diets on one hand had the same impact on the level of myofibre-myofibre detachments, and the EPA and DHA diets on the other hand. It thus seems that replacing fish oil (FO) by rapeseed oil (RO) in the diet of Atlantic salmon did not have any negative impact on muscle integrity at the myofibre-myofibre level. Salmon fillet texture is determined, among other things, by the level of myofibre-myofibre detachments (Taylor et al., 2002) and accordingly to our results, Regost et al. (2004) revealed no difference in texture measured instrumentally between Atlantic salmon fed an RO diet and a capelin oil diet, both diets being comparable to the RO and FO diets used in the present study.

In spite of the absence of significant differences between dietary groups at each time-point, during the whole storage period, the EPA diet caused significantly more myofibre-myofibre detachments than the FO diet ($p=0.022$) and to some extent, more than the RO diet ($p=0.073$). These results and the ones from Kjær et al. (2008) and Todorcevic et al. (2008) on the same salmon strongly suggest an increased oxidative stress in the EPA and DHA groups. Histological lesions of the muscle due to lipid peroxidation have already been observed (Brownell and Engel, 1978; García-Buñuel, 1984; Wada et al., 1991). Our results also agree with Izquierdo et al. (2005), who noticed a softer flesh texture in seabream fed diets with PUFA levels comparable to our extreme diets EPA and DHA. They attributed this softer texture to the higher amount of PUFAs and lower amount of saturated FAs in the diet/muscle of those fish, comparing to the control, as it was the case in our study.

b Myofibre-myocommata detachments, myofibre breakages and contracted myofibres

Microscopic observations of myofibre-myocommata detachments did not show any clear differences between dietary groups and no specific trend was observed through time *post-mortem* (results not shown). These results agreed with Taylor et al. (2002) and Ofstad et al. (2006) before 5 days in Atlantic salmon and 7 days in wolfish, as well as our previous study on super-chilling mentioned above.

No myofibre breakages were detected in longitudinal-section muscle samples 96 h *post-mortem* in any of the dietary groups.

Table 4 Average percentages of contracted myofibres in Atlantic salmon muscle at 96 h *post-mortem* ±SE. Different letters denote significant differences (p<0.05) between dietary groups (n=3 in each group).

Dietary groups	Contracted myofibre (%) ±SE
Fish oil (FO)	$14.07^{a}\pm5.05$
Rapeseed oil (RO)	$17.01^{a}\pm6.54$
EPA	$19.51^{a}\pm4.40$
DHA	$31.86^{b}\pm6.44$

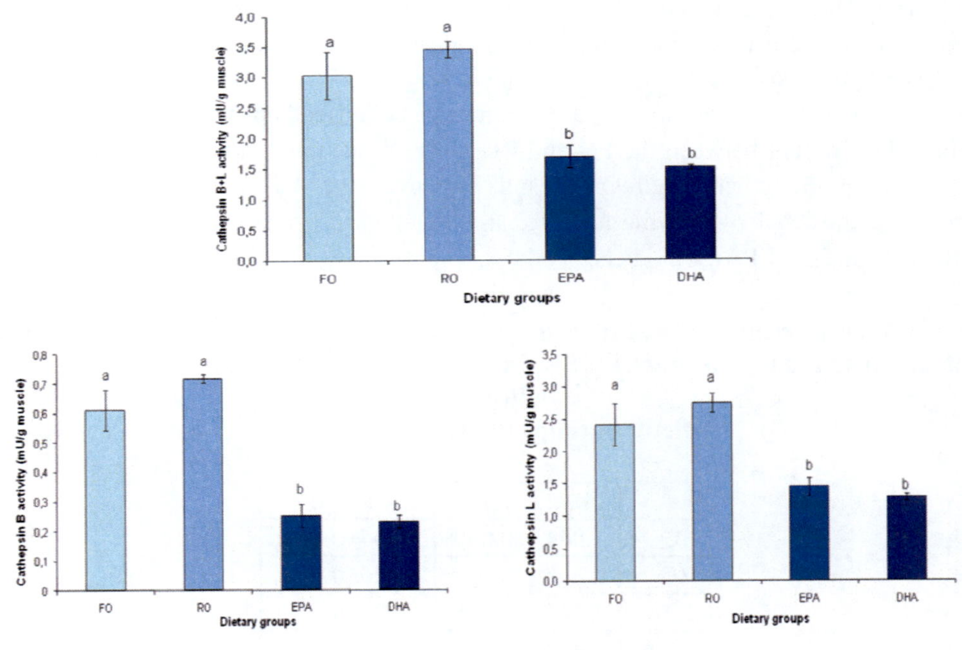

Figure 18. Average cathepsin B+L, cathepsin B and cathepsin L total activities in Atlantic salmon muscle homogenates, 0 h *post-mortem*. Each histobar represents the mean ± SE in each dietary group. Different letters denote significant differences (p<0.05) between the dietary groups (n=3 in each group) (Bahuaud et al., 2009b).

Table 4 presents the percentages of contracted myofibres in the four dietary groups, 96 h *post-mortem*, in longitudinal-section muscle samples. The DHA group presented significantly more contracted myofibres than the three other groups and feeding salmon with the DHA diet resulted in more than twice as much contracted myofibres as in the FO group (p=0.058). Mørkøre (2006) also showed a higher contraction rate of fillets of Atlantic cod fed a diet where fish oil was replaced at 40 % by soybean oil, compared to cod fed a 100 % fish oil diet. Mørkøre suggested that this difference might be due to accelerated energy depletion in the cod fed the soybean oil diet, since she found a linear relationship between *rigor* contraction and ATP depletion. She explained this faster ATP depletion by a reduced ability to cope with stress during slaughtering process or a different enzymatic activity for fish fed other lipid sources than fish oil. This last statement especially would agree with our results on cathepsins, as we found significant differences between the EPA/DHA groups and the FO/RO groups in cathepsin B+L, cathepsin B and cathepsin L activities as well as *cathepsin B* relative gene expression in muscle homogenates.

3.3.4. Cathepsin B+L, cathepsin B and cathepsin L total activities in muscle homogenates

Results from cathepsin B+L, cathepsin B and cathepsin L total activities in the 0 h *post-mortem* muscle homogenates are presented in Figure 18. For the four dietary groups, cathepsin L activity was found dominant when compared to cathepsin B activity, as we observed almost four times more cathepsin L than cathepsin B activity in the muscle homogenate. These results agree with Yamashita and Konagaya (1991), who showed that in salmon, cathepsin L was dominant in proteolytic degradation compared to cathepsin B, which had a lower hydrolysis rate.

Cathepsin B+L, as well as separated cathepsin B and cathepsins L activities were not significantly different between the FO and RO diets on one hand, and between the EPA and DHA diets on the other hand. Our results however showed two distinguished groups according to the level of enzyme activity: significant differences were found between the FO/RO diets and the EPA/DHA diets.

Table 5 Average relative gene expressions of *Cathepsin B* and *Cathepsin L* in Atlantic salmon muscle, 0 h *post-mortem* ±SE. Different letters denote significant differences between the dietary groups (n=3 in each group).

	Cathepsin B	*Cathepsin L*
FO	1.004^{ab}±0.069	1.004±0.071
RO	0.808^{a}±0.044	0.826±0.096
EPA	1.144^{bd}±0.158	0.998±0.118
DHA	1.432^{cd}±0.063	0.999±0.037

The EPA and DHA diets, considered as extreme diets due to their very high level of PUFAs, led to a reduction *in vivo* of cathepsin B and cathepsin L activities, when compared to the FO and RO diets. A lower cathepsin activity is usually associated with a lower potential for muscle degradation; less muscle degradation in the EPA and DHA groups than in the FO and RO groups would thus have been expected. These results on cathepsin activity did not agree with the percentage of myofibre-myofibre detachments found higher due to the EPA and to some extent to the DHA diets. The lower cathepsin B+L, as well as separated cathepsin B and cathepsin L activities found in salmon fed the EPA or DHA diet thus may result from an impairment of the lysosomal function or a disturbed biosynthesis of these enzymes (Kharbanda et al., 1995 and 1996), here possibly due to oxidative stress (Sargent et al., 2002). Headlam et al. (2006) also showed that cathepsins are inactivated by protein hydroperoxides. In the present study, production of hydroperoxides due to potential oxidative stress in the EPA and DHA fed salmon, could thus explain the lower cathepsin B and cathepsin L activities in the muscle.

Consequently, cathepsin B and cathepsin L did not seem to be directly responsible for the observed muscle degradation, as their total activity did not correlate with the level of myofibre-myofibre detachments. This underlines the complexity in muscle degrading processes and the necessity for further research in this area. In the present study, other protease activity as calpains may have been involved (Delbarre-Ladrat et al., 2004a and 2004b; Chéret et al., 2007). Results from the same experiment as the present one, on mitochondrial function from liver and adipose tissue of the same salmon (Kjær et al., 2008; Todorcevic et al., 2008), do, however indicate that lipid peroxidation and related protein denaturation should also be considered.

3.3.5. Cathepsin B and cathepsin L relative gene expressions

Relative gene expressions of *cathepsin B* and *cathepsin L* are presented in Table 5. Only the expression of *cathepsin B* was significantly influenced by the diet (p=0.016), whereas the expression of *cathepsin L* was not significantly different between dietary groups.

The unexpected effects of dietary lipids on the *in vivo* level of cathepsin B+L, as well as separated cathepsin B and cathepsin L activities obtained in the present study demonstrate an impact of dietary lipid composition on lysosomal proteases. Alizadeh et al. (2006) showed that oxidative stress increases mRNA expression of *cathepsin B* and *cathepsin L* in the retinal pigment epithelium of mouse, with consequent degeneration of the retina. Accordingly and in agreement with our results discussed above, oxidative stress could thus be responsible for the increased relative gene expression of *cathepsin B* found in the EPA and DHA groups in our study. This higher gene expression would then be the response to an increased need to get rid of more damaged materials due to oxidative stress.

In rainbow trout, Salem et al. (2006) showed that during spawning and post-spawning, the relative gene expression of *cathepsin D* was not correlated with the relative activity of cathepsin D, but that the relative gene expression of *cathepsin L* was correlated with the relative activity of cathepsin L. Our gene expression study did not correlate with the lower cathepsin activity in the EPA and DHA groups. It thus seems that regulation of cathepsin activity is a complex system and further investigations are needed.

3.4. Study III: Pre-Slaughter Crowding Stress (Bahuaud Et Al., 2009a)

3.4.1. pH of pre-rigor and post-rigor fillets

Muscle pH significantly decreased with time of storage (p<0.0001) in the *pre-rigor* fillets (Figure 19) and over the whole storage period of the *pre-rigor* fillets, a significant effect of pre-slaughter stress on muscle pH was seen in general (p=0.0003); the LS group presenting the lowest muscle pH values ($p_{(NS/LS)}$=0.002; $p_{(SS/LS)}$=0.0001). Our results agree with other studies showing a lower muscle pH immediately *post-mortem* or a higher lactate level in stressed fish than in unstressed ones (Bagni et al., 2007; Erikson et al., 1997a; Erikson and Misimi, 2008; Lowe et al., 1993; Stien et al., 2005; Tejada and Huidobro, 2002; Thomas et al., 1999). Pre-slaughter stress accelerates anaerobic energy metabolism, and therefore produces a rapid increase in muscle lactic acid production (Poli et al., 2005), leading to a lowering of the muscle pH. The strong pH reduction in long-term stressed salmon immediately *post-mortem* may suggest an *in vivo* accumulation of lactic acid in the muscle, because long-term stress induced strong swimming activity and presumably a very slow release of lactic acid from the cells.

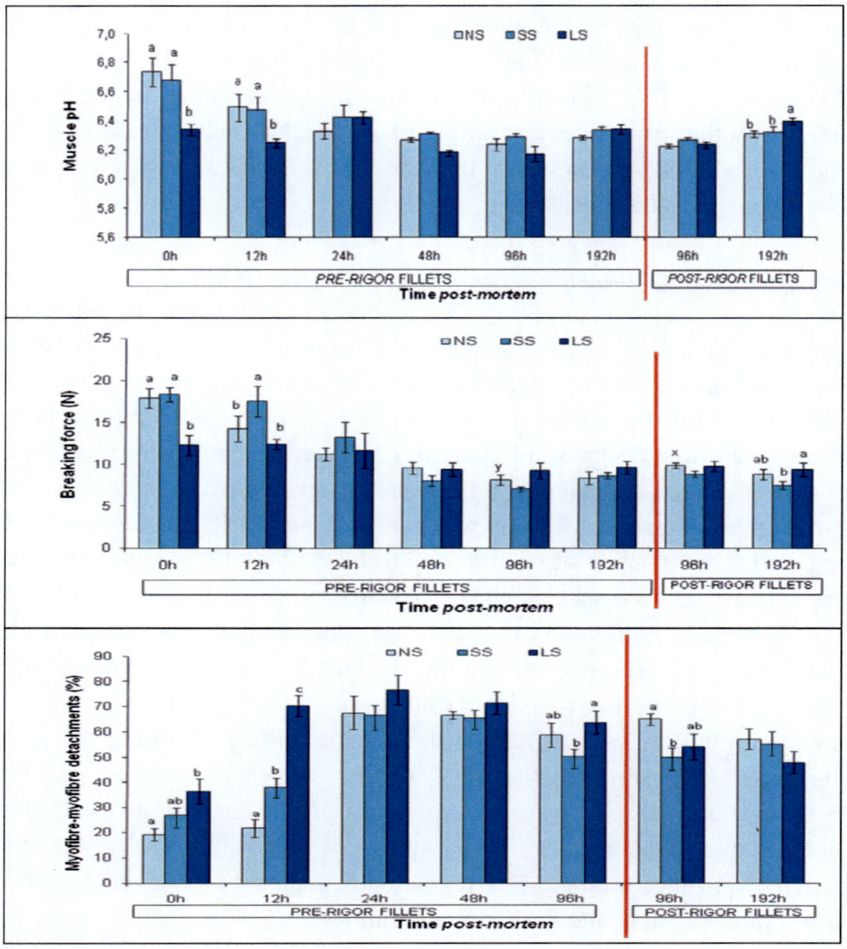

Figure 19. Average pH, breaking force (fillet firmness), percentages of myofibre-myofibre detachments and percentages of myofibre-myocommata detachments in Atlantic salmon *pre-rigor* and *post-rigor* fillets subjected to Minimal stress (NS), Short-term Stress (SS) or Long-term Stress (LS). Each histobar represents the mean ±SE in each stress groups at each time-points *post-mortem*. Different letters denote significant differences ($p<0.05$) between stress groups (n=6 in each group) at the same time-point (Bahuaud et al., 2009a).

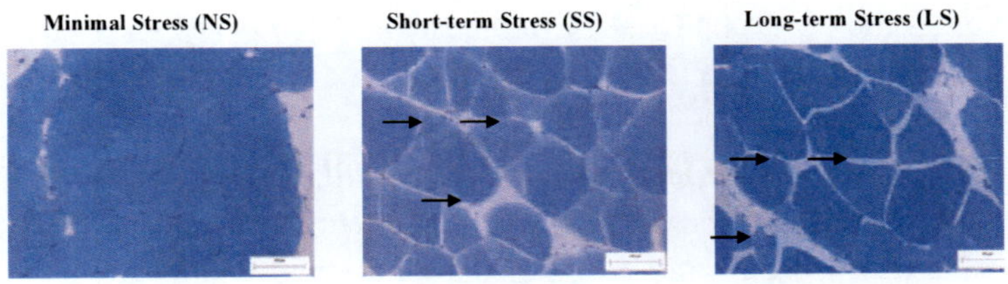

Figure 20. Light microscopy observations (x20) 0 h *post-mortem*, of *pre-rigor* fillets of Atlantic salmon subjected to minimal (NS), short- (SS) or long-term (LS) pre-slaughter crowding stress. Arrows: myofibre-myofibre detachments (Bahuaud et al., 2009a).

In the *post-rigor* fillets, pH significantly changed between 96 h and 192 h *post-mortem* ($p<0.0001$) but no significant effect of pre-slaughter stress was found in general during this period. At 192 h *post-mortem*, fillets from the LS group however presented the highest pH ($p_{(NS/LS)}=0.019$ and $p_{(SS/LS)}=0.043$).

Muscle pH values showed no significant difference between *pre-* and *post-rigor* fillets at 96 h and 192 h *post-mortem*.

3.4.2. Fillet firmness of pre-rigor and post-rigor fillets

Fillet firmness significantly changed with time of storage ($p<0.0001$) in the *pre-rigor* fillets (Figure 19). Significant differences in fillet firmness between stress groups were seen at the beginning of the storage period, but not after 24 h *post-mortem*; and this agrees with Erikson et al. (1997a), Skjervold et al. (2001a) and Stien et al. (2005), who noticed that pre-slaughter stress had the strongest impact at early stages *post-mortem* in Atlantic salmon and cod, but gradually became less pronounced. Over the whole storage period of the *pre-rigor* fillets, there was a tendency that pre-slaughter crowding stress had an effect on fillet texture (F_{break}) ($p=0.087$) and long-term stress (LS) had a significantly different impact than short-term stress (SS) ($p=0.029$). Our results on fillet firmness agree with Ando et al. (2001), who also found that the breaking strength of stressed chub mackerel was lower than in non-stressed ones at early stages *post-mortem*. Like in our study, this softer texture was associated with a higher incidence of myofibre-myofibre detachments. Godiksen et al. (2009) showed that cathepsin B, D and L were also correlated to fillet texture in rainbow trout, and these results would agree with the ones obtained in our study on cathepsin B activity and *cathepsin L* gene expression.

In the *post-rigor* fillets, fillet firmness significantly decreased between 96 h and 192 h *post-mortem* ($p=0.051$). There was a significant effect of pre-slaughter stress on fillet texture during this storage period in the *post-rigor* fillets ($p=0.041$), with the SS group presenting the softest fillets significantly different from the NS group ($p=0.046$). At 192 h *post-mortem*, the SS fillets were also softer than the LS fillets ($p=0.028$). These results are not easily understood, but agree with Kiessling et al. (2004) who found that short-term stressed Atlantic salmon presented a softer muscle texture after around 1 week of storage.

At 96 h and 192 h *post-mortem*, time of filleting did not have any significant influence on fillet texture in general, but 96 h *post-mortem*, *pre-rigor* fillets from the NS group were significantly softer ($p=0.054$) than *post-rigor* fillets from the same group. The same tendency could be observed in the SS group (p=0.080), but these differences had disappeared by 192 h *post-mortem*.

3.4.3. Myofibre-myofibre detachments in pre-rigor and post-rigor fillets

In the *pre-rigor* fillets, the percentage of myofibre-myofibre detachments significantly increased with time of storage (p<0.0001), translating the normal muscle degradation process during storage (Figure 19). Detachments also increased because of pre-slaughter crowding stress (p<0.0001) (Figures 19 and 20). As for breaking force (fillet firmness), significant differences in percentages of myofibre-myofibre detachments between groups at each time-point were mainly noticed before 24 h *post-mortem*. But over the whole storage period, the LS group showed the highest percentage of myofibre-myofibre detachments ($p_{(NS/LS)}$<0.0001 and $p_{(SS/LS)}$<0.0001) in the *pre-rigor* fillets. These results agree with Ando et al. (2001) and Nakayama et al. (1996), who also showed that pre-slaughter stress increased the incidence of myofibre-myofibre detachments in chub mackerel and red sea-bream.

Ofstad et al. (2006) showed that both myofibre-myofibre and myofibre-myocommata detachments in cod and wolfish were mainly caused by fractures in the extracellular matrix, maybe due to degradation of proteoglycans and glycoproteins. Besides, increase of plasma cortisol is a well-known primary stress response in fish, and in accordance with our results, Macdonald and Cidlowski (1982) showed that glucocorticoids in rat stimulate protein degradation. The acceleration of detachments between myofibres due to pre-slaughter stress and observed particularly in the LS group in the present study, could also be related to the significantly lower muscle pH immediately *post-mortem*, since a significant negative correlation was found between muscle pH and percentage of myofibre-myofibre detachments.

In the *post-rigor* fillets, no significant difference in percentage of myofibre-myofibre detachments was noticed in general between 96 h and 192 h *post-mortem* or between stress groups. But 96 h *post-mortem,* as in the *pre-rigor* fillets, short-term stress (SS) resulted in less myofibre-myofibre detachments, significantly when compared to the NS group (p=0.040). This difference had disappeared by 192 h *post-mortem*.

At 96 h *post-mortem*, no significant differences in percentage of myofibre-myofibre detachments were found between the *pre-rigor* and the *post-rigor* fillets.

3.4.4. Myofibre-myocommata detachments in pre-rigor and post-rigor fillets

No general effect of storage time was observed between 0 h and 96 h *post-mortem* in the *pre-rigor* fillets (Figure 19), but some significant individual increases between some time-points could be noticed. Pre-slaughter stress further increased the level of myofibre-myocommata detachments in general (p=0.047) over the whole storage period in the *pre-rigor* fillets, and significantly more myofibre-myocommata detachments were found in the SS (p=0.028) and in the LS groups (p=0.037) than in the NS group. Accordingly, Gomez-Guillen et al. (2000) found that shear strength of the connective tissue fraction in stressed Atlantic salmon was lower, and this was associated with lower muscle pH. In earlier studies including ours on super-chilling and dietary lipids described above, such increase in myofibre-myocommata detachments was not noticed until 192 h *post-mortem* in Atlantic salmon and wolfish (Bahuaud et al., 2008; Bahuaud et al., 2009b; Ofstad et al., 2006; Taylor et al., 2002) and it was suggested that myofibre-myocommata detachments were associated with loss of *rigor* stiffness (Taylor et al., 2002). In the present study, the very low muscle pH found particularly in the LS group could be directly responsible for these detachments, as we also found a significant negative correlation between muscle pH and percentage of myofibre-myocommata detachments. Lavety et al. (1988) also showed a strong negative relationship between muscle pH and degree of gaping in salmon and rainbow trout, and also noticed that the lower the muscle pH, the easier to break the myocommata.

Table 6. Average percentages of myofibre breakages and contracted myofibres in Atlantic salmon *pre-rigor* and *post-rigor* fillets in the different stress groups (NS, SS and LS), 96 h *post-mortem* ± SE. Different letters denote significant differences (*p*<0.05) between the stress groups (a, b, c) (n=6 in each group) or between times of filleting (*pre-/post-rigor*) (x, y).

	Myofibre breakages (%)		Contracted myofibres (%)	
	Pre-rigor fillets	*Post-rigor fillets*	*Pre-rigor fillets*	*Post-rigor fillets*
NS	14.42^{b}±1.85	8.99±2.46	14.63±2.57	14.56^{b}±3.37
SS	19.95^{b}±3.49	14.58±1.79	22.13±5.04	28.83^{a}±5.15
LS	23.02^{ax}±0.89	11.20^{y}±1.72	16.68^{y}±2.68	29.99^{ax}±5.09

The level of myofibre-myocommata detachments significantly increased between 96 h and 192 h *post-mortem* (p=0.002) in the *post-rigor* fillets. Like in the *pre-rigor* fillets, the SS group showed significantly more myofibre-myocommata detachments than the NS group (p=0.035) but this time only after 192 h of storage.

At 96 h *post-mortem*, *pre-rigor* filleting resulted in significantly more myofibre-myocommata detachments than *post-rigor* filleting, but only when salmon were subjected to short-term (SS) stress (p=0.034).

3.4.5. Myofibre breakages and contracted myofibres in pre-rigor and post-rigor fillets

In the *pre-rigor* fillets, significantly more myofibre breakages were found in the long-term-stressed salmon (Table 6). Besides, salmon subjected to long-term stress (LS) presented the highest fillet contraction (results not shown) (Martinsen, 2006). The higher percentage of myofibre breakages observed in the stressed salmon could thus result from the higher tension exerted on the fillet during *rigor* contraction (Erikson et al., 1999; Erikson and Misimi, 2008; Nakayama et al., 1994; Skjervold et al., 1999; Skjervold et al., 2001a; Stien et al., 2005; Thomas et al., 1999; Veiseth et al., 2006). Taylor and Koohmaraie (1998) found more breakages in the I-band region of myofibrils in normal sheep muscle than in callipyge, where callipyge muscle was associated with less proteolytic activity. In our study, the higher cathepsin B activity in the stressed salmon 0 h *post-mortem*, even not significant, could thus also be responsible for the higher percentage of myofibre breakages observed in these fish (significantly in the LS group), when compared to the NS group.

In the *post-rigor* fillets, no significant differences in percentage of myofibre breakages were noticed between the three groups.

Pre-rigor filleting resulted in significantly more myofibre breakages than *post-rigor* filleting in general (p=0.003), and especially when salmon were subjected to long-term stress (LS).

In the *pre-rigor* fillets, pre-slaughter stress did not influence the percentage of contracted myofibres, whereas in the *post-rigor* fillets, short-term (SS) (p=0.058) and long-term pre-slaughter stress (LS) (p=0.072) tended to result in more contracted myofibres than minimal stress (Table 6). There was also a tendency that *pre-rigor* filleting resulted in less contracted myofibres than *post-rigor* filleting (p=0.063), and this difference was significant for salmon subjected to long-term stress (LS). These results are in accordance with several authors (Erikson and Misimi, 2008; Erikson et al., 1999; Nakayama et al., 1994; Skjervold et al., 1999 and 2001a; Stien et al., 2005; Thomas et al., 1999; Veiseth et al., 2006) who pointed out that pre-slaughter stress enhances *rigor* contraction in fish, probably due to a more rapid depletion of ATP.

3.4.6. Cathepsin activity and relative gene expression in pre-rigor fillets

Pre-slaughter crowding stress (SS and LS) seemed to result, 0 h *post-mortem*, in more cathepsin B+L total activity in the muscle than minimal crowding stress (NS), even though these differences were not significant ($p_{(NS/LS)}$=0.141) (Figure 21). Exactly the same pattern was observed when measuring cathepsin B activity alone ($p_{(NS/LS)}$=0.084), whereas cathepsin L activity did not seem to be affected by pre-slaughter crowding stress.

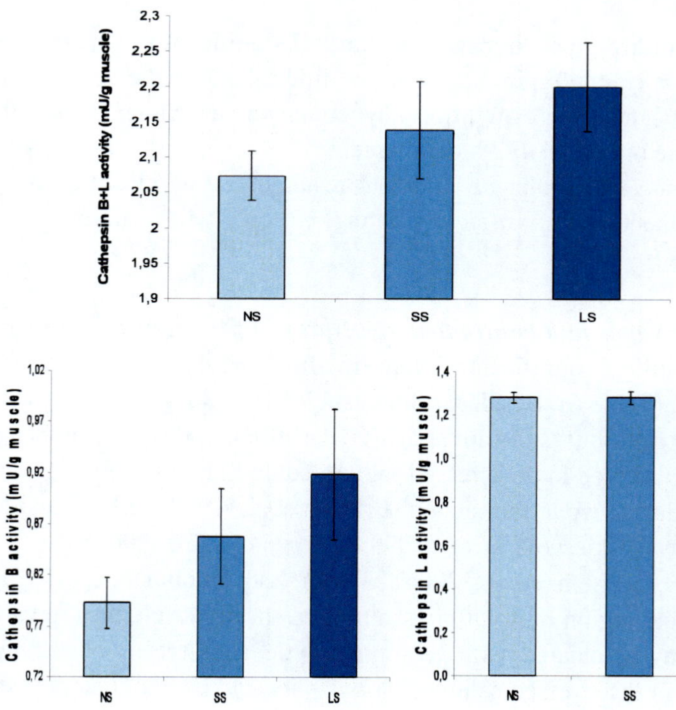

Figure 21. Average Cathepsin B+L, cathepsin B and cathepsin L total activities, 0 h *post-mortem*, in muscle homogenates of Atlantic salmon subjected to pre-slaughter Minimal Stress (NS), Short-term Stress (SS) or Long-term Stress (LS). Each histobar represents the mean value ±SE in each stress group (n=6 in each group) (Bahuaud et al., 2009a).

The relative gene expression of *cathepsin L* was significantly influenced by pre-slaughter stress (p=0.001) 0 h *post-mortem*, with the LS group presenting a significantly higher expression than the two other groups (Figure 22). There was a tendency that the gene expression of *cathepsin B* followed the same pattern, even if these results were not significant (Table 7).

These results agree with studies on rats and mice where, under stress conditions, cathepsin total activity (Pote and Altekar, 1980; Vihko and Salminen, 1983) increased in the muscle. In salmonids, higher cathepsin B and cathepsin L have been particularly associated with muscle softening, with an optimum pH for activity of 6.0-5.5 and 5.6 respectively (Aoki and Ueno, 1997; Aoki et al., 1995; Godiksen et al., 2009; Yamashita and Konagaya, 1990; Yamashita and Konagaya, 1991). In the present study, we found a significant negative correlation between muscle pH and level of cathepsin B+L activity. Pre-slaughter stress resulted in lower muscle pH immediately *post-mortem* than minimal-stress (NS), significantly for long-term stress (LS), and this low pH could be directly responsible for the earlier higher activity of cathepsin B in the muscle of the stressed salmon, especially the LS group. Gil et al. (1999) also measured a higher activity of cathepsin B and cathepsin B+L in PSE (Pale Soft Exudative) ham with low pH, when compared with normal quality ham. Accordingly, in mammals, Taylor et al. (1995) suggested that cathepsins participated in mammalian muscle proteolysis *post-mortem* as soon as muscle pH became low.

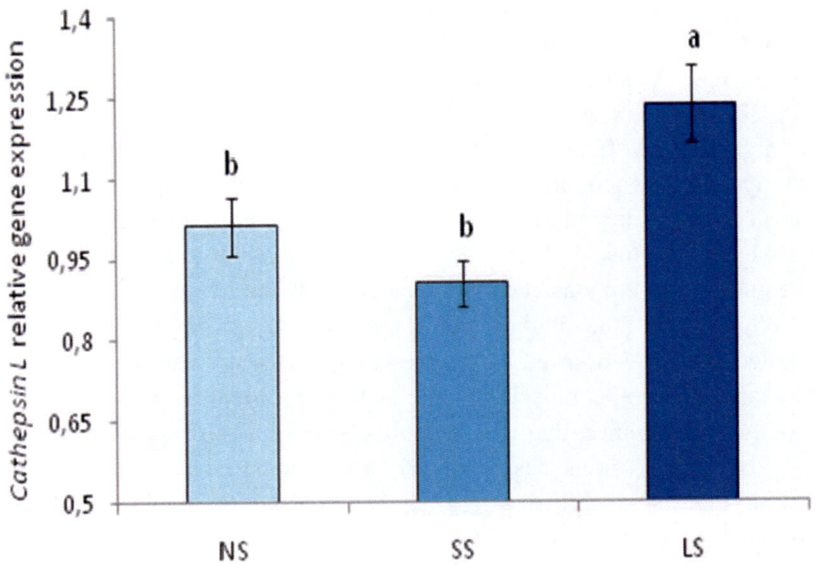

Figure 22. Average *cathepsin L* relative gene expression 0h *post-mortem*, in muscle homogenates of Atlantic salmon subjected to Minimal Stress (NS), Short-term Stress (SS) or Long-term Stress (LS). Each histobar represents the mean ±SE for each stress groups. Different letters denote significant differences (p<0.05) between stress groups (n=6 in each group).

Table 7. Average *cathepsin B* relative gene expression 0h *post-mortem*, in muscle homogenates of Atlantic salmon subjected to Minimal Stress (NS), Short-term Stress (SS) or Long-term Stress (LS) ±SE.

	Cathepsin B relative gene expression
NS	1.063±0.115
SS	1.014±0.043
LS	1.245±0.107

3.4.7. Correlations between the quality traits and muscle degradation parameters in pre-rigor and post-rigor fillets

Table 8 summarizes the correlation coefficients between the results obtained from the different parameters analysed in the present study on stress. Significant positive correlations between percentage of myofibre-myofibre detachments and cathepsin B activity on one hand, and *cathepsin L* gene expression on the other hand, were observed, as well as between percentage of myofibre-myocommata detachments and cathepsin B activity. These results agree with Vihko and Salminen (1983), who found an increase in cathepsin D activity in exhausted rat muscle, and this activity increase was associated with muscle fibre injuries, necrosis and protein degradation. Gomez-Guillen et al. (2000) also noticed that pre-slaughter stress in Atlantic salmon resulted in myofibrillar protein denaturation and greater myosin proteolysis, knowing that cathepsins are able to degrade myosin easily (Aoki and Ueno, 1997; Ladrat et al., 2003; Taylor et al., 1995). In the present study, the significantly higher percentage of myofibre-myofibre detachments until 24 h *post-mortem* and significantly higher percentage of myofibre-myocommata detachments over the whole storage period found in the stressed groups could thus be a direct consequence of the action of cathepsin B activity in the muscle, the lower initial muscle pH *post-mortem* in these groups, especially in the LS group, allowing the action of this enzyme at earlier stages than in the NS group.

On the other hand, the higher percentage of myofibre-myofibre and myofibre-myocommata detachments observed in the present study could also be due to a mechanical stress directly applied to the muscle by pre-slaughter crowding stress. Muscle structure degradation of Atlantic salmon due to mechanical stress has already been suggested in our previous study on dietary lipids described above (Bahuaud et al., 2009b) and Roth et al. (2006) indicated that pre-slaughter stress in Atlantic salmon could affect myofibrils or connective tissue physically, thus releasing proteases and accelerating muscle degradation. Gomez-Guillen et al. (2000), Sigholt et al. (1997) and Yamashita and Konagaya (1991) also suggested that protein denaturation resulting from exposure of *post-mortem* muscle to low pHs, may facilitate the access of substrates by proteolytic enzymes and thus enhancing muscle proteolysis. The low muscle pH observed especially in the long stressed (LS) salmon could also have caused a direct damage of the muscle structure by shrinking myofibres, thus increasing the tensions on the endomysium or at the connective tissue level, resulting in a higher number of myofibre-myofibre and myofibre-myocommata detachments.

Table 8. Correlation coefficients between results of the different quality traits (Bahuaud et al., 2009a)

	pH	Myof-myof detach	Myof-myoc detach	Firmness	Cath B+L act	Cath B act	Cath L act	*Cath B* exp	*Cath L* exp
pH		-0.548***	-0.222**	0.292***	-0.460**	-	-	-	-
Myof-myof detach	-0.548***		-	-0.516***	-	0.477**	-	-	0.513**
Myof-myoc detach	-0.222**	-		-0.204**	*-0.435**	0.455**	-	-	-
Firmness	0.292***	-0.516***	-0.204**		-	*-0.437**	-	-	*-0.469**
Cath B+L activity	-0.460**	-	*0.435**	-		nd	nd	nd	nd
Cath B activity	-	0.477**	0.455**	*-0.437**	nd		nd	-	nd
Cath L activity	-	-	-	-	nd	nd		nd	-
Cath B expression	-	-	-	-	nd	-	nd		nd
Cath L expression	-	0.513**	-	*-0.469**	nd	nd	-	nd	

* Correlation trend p<0.07; ** Correlation is significant at p<0.05; *** Correlation is significant at p<0.001; negative signs denote no significant correlation; nd=correlation not-determined

Table 9. Summary of the results and hypothetical causes of the lysosomal system activation and consequent quality degradation observed in the three studies (I, II and III). nm: not measured; LS: long-term stress; SS: short-term stress; empty space means no influence.

	BIOCHEMICAL FACTORS		
"QUALITY" ANALYSES	*Super-chilling -1.5 °C*	*>50 % dietary n-3 fatty acids (EPA and DHA)*	*Pre-slaughter crowding stress*
Study	I	II	III
Hypothetical causes of lysosomal system activation and consequent quality degradation	Formation of ice crystals	High oxidative stress	Low muscle pH already *pre-mortem* and immediately *post-mortem* (LS group)
Fatty acid composition of the lysosomes	nm	Increase percentage of PUFAs: increase fragility of the membrane	nm
Cathepsin B and L activity in the lysosomes during storage	Acceleration of decrease level	nm	nm
Cathepsin B and L total activity in the muscle homogenate		Decreases (probably due to inhibition by hydroperoxides)	Cathepsin B increases (tendency)
Cathepsin B and L relative gene expression in the muscle	nm	Cathepsin B increases	LS increases cathepsin L
Myofibre-myofibre detachments	Accelerated until 24 h *post-mortem*	Accelerated until 24 h *post-mortem* (tendency)	Accelerated until 24 h *post-mortem*
Myofibre-myocommata detachments			Increase, especially with SS
Myofibre breakages	Increase		Increase (in *pre-rigor* fillets)
Liquid leakage	Increase	nm	nm
Contracted myofibres	Decrease	Increase in DHA group	Increase (in *post-rigor* fillets)
Fillet firmness		nm	LS accelerated softness until 24 h *post-mortem*

In the present study on stress, a strong significant correlation between muscle pH and fillet softness (F_{break}) was also observed, similar to the one suggested by Stien et al. (2005) on cod. This could indicate that the lower initial muscle pH as revealed in the LS group was responsible for muscle structure degradation, influencing fillet texture, as suggested by Morzel et al. (2003) in turbot. There were also significant correlations between muscle texture and percentage of myofibre-myofibre detachments, myofibre-myocommata detachments, and *cathepsin L* gene expression. Fillet texture thus resulted from a combination of different factors as already suggested by Skjervold et al. (2001b) and Taylor et al. (2002); pre-slaughter stress causing a lower initial muscle pH, activating cathepsins earlier, and these cathepsins acting directly on muscle integrity and influencing fillet texture. In the present study, the attention has been given to the lysosomal proteases cathepsins, as these enzymes have previously been involved in *post-mortem* degradation of fish, and we established a possible link between low pH due to pre-slaughter stress and cathepsin activity. In addition, the eventual participation of other proteases like calpains cannot be excluded and investigations on these cytosolic enzymes are currently being processed in our laboratory.

The influence of time of filleting (*pre-rigor* or *post-rigor*) was compared at 96 h and 192 h *post-mortem* and little impact on the observed quality traits and muscle degradation parameters was found. Our results are in accordance with Skjervold et al. (2001b), who noticed that quality differences between *pre-rigor* and *post-rigor* fillets are reduced with time. We found, however, that *pre-rigor* filleting increased the percentage of myofibre breakages 96 h *post-mortem*. *Pre-rigor* fillets are known to contract more than *post-rigor* fillets, as in *pre-rigor* fillets, *rigor* contraction can progress without resistance from the fish backbone (Mørkøre et al., 2008; Skjervold et al., 2001b; Skjervold et al., 2001c). This higher *rigor* contraction would then increase tensions in the myofibres, resulting in a higher number of breakages. Misimi et al. (2008) and Mørkøre et al. (2008) also suggested that the effects of *ante-mortem* treatments (here pre-slaughter stress) are more pronounced in *pre-rigor* than *post-rigor* fillets, as significant effects of pre-slaughter stress usually occur before a noticeable onset of *rigor mortis*. *Post-rigor* filleting tending to increase the percentage of contracted myofibres 96 h *post-mortem* was however difficult to explain and therefore further investigation is necessary to understand this phenomenon.

3.5. Summary of the three Studies and General Reflexions on Salmon Muscle Degradation

3.6. Effects of Super-Chilling

In study I (Bahuaud et al., 2008), super-chilling of Atlantic salmon leading to a fillet core temperature of -1.5 °C resulted in the formation of intra- and extra-cellular ice crystals in the upper layer of the fillets, probably being the cause of the faster release of lysosomal cathepsins into the cytosol. The release of these enzymes as a probable consequence of lysosomal breakages already started 12 h after treatment, and this release was continuing during the whole storage period. Increase of myofibre-myofibre detachments paralleled the decrease of cathepsin B + L in the lysosomal fraction (pellet) during storage. The percentage of myofibre breakages and liquid leakage 144 h after treatment was higher in the super-chilled fillets, translating a possible relationship between these two phenomena. Ice crystals thus seemed to cause myofibre breakages, either mechanically, or by the action of lysosomal cathepsins, and these damages in the muscle fibres may be responsible for the increase loss of liquid leakage observed in the super-chilled fillets. We concluded that in spite of the numerous advantages that the technique could bring to salmon consumption, the technical parameters of the super-chilling process need to be optimized in order to avoid the formation of ice crystals, and the following acceleration of salmon muscle degradation.

3.7. Effects of Dietary N-6 and N-3 Fatty Acids

Study II (Bahuaud et al., 2009b) described the effects of dietary n-6 fatty acids from rapeseed oil and very high levels of n-3 fatty acids (EPA and DHA) on the lysosomal system, and muscle integrity. Total replacement of fish oil (FO) by rapeseed oil (RO) in Atlantic salmon feed did not lead to any significant changes when it comes to muscle integrity (myofibre-myofibre and myofibre-myocommata detachments), muscle cathepsin B and L total activities or gene expressions. Very high levels (> 50 %) of n-3 fatty acids (EPA and DHA) in the diet however resulted in a higher muscle degradation evaluated by a higher percentage of myofibre-myofibre detachments, already 0 h *post-mortem* even if these results were not significant. These results could be linked to the ones of Taylor et al. (1997), who also strongly suggested a participation of lipids in muscle cytoskeleton stability. These two extreme diets probably lead to an impairment of the lysosomal system or cathepsin biosynthesis due to a high oxidative stress in the muscle (Kharbanda et al., 1995 and 1996; Sargent et al., 2002), as shown by parallel analyses on mitochondria and peroxisomes in other studies but on the same fish (Kjær et al., 2008; Todorcevic et al., 2008). Oxidative stress results in formation of hydroperoxides, inhibiting cathepsin activity (Headlam et al., 2006). Therefore, Atlantic salmon fed with very high levels of n-3 fatty acids showed reduced muscle cathepsin B + L total activity immediately *post-mortem*. *Cathepsin B* gene expression was, however, higher in these two extreme groups, significantly in the DHA group, than in the (FO) and RO fed salmon. These results indicated that cathepsin B was influenced and the lysosomal function probably impaired by such dietary treatments. Fatty acid composition of the lysosomal fraction depended of the fatty acid composition of the diet. While the FO and RO groups did not present any significant differences in lysosomal fatty acid composition in general, the two extreme diets (EPA and DHA) showed a very high level of PUFAs in the muscle lysosomal fraction, possibly leading to a fragility of the lysosomal membrane.

The lesions observed at the muscle structure level in this case may be due to the potential high lipid peroxidation, phenomenon already reported by Brownell and Engel (1978), Garcia-Bunuel (1984) and Wada et al. (1991). Izquierdo et al. (2005) also noticed that comparable diets containing very high levels of PUFAs increased muscle softness in sea bream, because of the lower stability of PUFAs compared to SFAs. It is however not excluded that the degradation of muscle quality observed in this study could also have a connection with the higher *cathepsin B* gene expression, but this hypothesis needs more investigations.

3.8. Effects of Pre-Slaughter Crowding Stress

In study III (Bahuaud et al., 2009a), the effects of pre-slaughter crowding stress on Atlantic salmon muscle integrity and on the potential roles of the lysosomal system and its catheptic enzymes were investigated. Pre-slaughter long-term crowding stress (LS) (24 h), resulted in a significant reduction of muscle pH immediately *post-mortem*, due to the production of lactic acid probably already *ante-mortem*. The optimum pH for cathepsin B and L activities ranges around 6 (Aoki et al., 1995; Aoki and Ueno, 1997; Godiksen et al., 2009; Yamashita and Konagaya, 1990 and 1991). The low pH in the LS group apparently started the activity of muscle cathepsin B + L earlier than when the pH was higher (in NS and SS groups), as cathepsin B + L activity was found increased in the LS salmon. Pre-slaughter stress also enhanced the percentages of myofibre-myofibre and myofibre-myocommata detachments in Atlantic salmon fillets. The higher cathepsin B + L activity may have directly influenced muscle quality by accelerating myofibre-myofibre detachments and increasing myofibre-myocommata detachments, as cathepsins can act at the connective tissue level (Yamashita and Konagaya, 1991). Muscle softening was accelerated in the LS salmon, a possible consequence of the higher cathepsin B + L activity and the higher *cathepsin L* gene expression in this group (Godiksen et al., 2009; Yamashita and Konagaya, 1991). Muscle pH was significantly correlated to myofibre-myofibre and myofibre-myocommata detachments, fillet firmness and cathepsin B + L total activity: low muscle pH may be responsible for the activation of cathepsin B + L in the muscle, and consequently for muscle structure degradation and fillet softness, as cathepsin B + L activity was also significantly correlated to myofibre-myofibre detachments.

3.9. Potential Importance and Role of Lysosomal Cathepsins B and L in Degradation of Atlantic Salmon Muscle Structure

When measured in the lysosomal fraction (pellet), cathepsin B + L activity decreased with storage time (study I), and this decrease paralleled the degradation of muscle structure (myofibre-myofibre and sometimes myofibre-myocommata detachments). When measured as total activity and/or gene expression in muscle homogenates, an increase in cathepsin corresponded to an increase in muscle degradation (studies II and III), evaluated as percentage of myofibre-myofibre and myofibre-myocommata detachments, myofibre breakages and/or fillet texture. This agrees with Béchet et al. (2005), who reported that the activation of the lysosomal system with increased expression of cathepsins appears as an early response to myofibrillar damage. Ofstad et al. (2006) also suggested that both myofibre-myofibre and myofibre-myocommata detachments in cod and wolfish were mainly due to fractures in the extracellular matrix, caused by the degradation of proteoglycans and other glycoproteins, whose degradation depends on several lysosomal enzymes (Junqueira et al., 1989b).

Cathepsin B and L activities and/or gene expressions were influenced by the three main factors studied here (super-chilling, dietary lipids and pre-slaughter crowding stress). Our results indicated that the lysosomal system including catheptic proteases could play an important role in muscle degradation and quality, both *pre-mortem* and *post-mortem*. Cathepsins could exert their activity on connective tissue (attachments between myofibres and between myofibres and myocommata) and eventually influence fillet texture (study III). Super-chilling formed destructive ice crystals in the muscle cells and increased the percentage of myofibre breakages (study I); these phenomena paralleling the release of cathepsins from the lysosomes to the cytosol. These results could highlight the impact of cathepsins also at the myofibrillar level, as cathepsins B and L have shown to be able to degrade myofibrillar proteins (Ladrat et al., 2003; Yamashita and Konagaya, 1991 and 1990), but further investigations are needed to confirm this hypothesis in the case of super-chilling.

Cathepsins seem to be strongly influenced by muscle pH, and the optimum pH for cathepsins B and L activities ranges around 6 (Aoki et al., 1995; Aoki and Ueno, 1997; Godiksen et al., 2009; Yamashita and Konagaya, 1990 and 1991). In study III, the low muscle pH due to long-term pre-slaughter stress seemed to activate cathepsin activity in the muscle, especially cathepsin B, possibly being responsible for the negative effects on muscle integrity and fillet texture. It is therefore important to control muscle pH *in vivo* and immediately after slaughter by reducing pre-slaughter stress. Feeding status before slaughter might have the same impact; as starved fish would use their glycogen reserves for energy, resulting in the production of lactic acid in the muscle and consequent low muscle pH already *in vivo* and immediately *post-mortem*.

The lysosomal catheptic system seems to act in a complex way and could be impaired by oxidative stress, as muscle *cathepsin B* gene expression was increased but cathepsin B + L total activity was reduced in salmon fed very high levels of n-3 fatty acids compared to control salmon (study II). This agrees with Tardy et al. (2006) who reported that oxidative stress can induce very fast lysosomal disruption creating intralysosomal redox reactions. Cathepsin B + L activity, in this case, was probably inhibited by hydroperoxides formed during oxidative stress (Headlam et al., 2006), and muscle degradation could thus be due to either mechanical stress on the myofibre-myofibre attachments and myocommata, or the action of other degrading proteases like calpains or collagenases (Salem et al., 2006; Sovik and Rustad, 2006; Suarez et al., 2005). The fact that *cathepsin B* relative gene expression was significantly increased in the very high level of n-3 fatty acid groups (EPA and DHA) could also mean the implication of cathepsin B in muscle degradation, but the mechanisms behind its action needs to be investigated.

In studies II and III, *cathepsin B* and *L* relative gene expressions were not always correlated with cathepsin B and L activities in the muscle, which agrees with Béchet et al. (2005), who reported that cathepsin mRNA levels are not systematically related to lysosomal cathepsin activity and Salem et al. (2006), who showed that relative gene expression of *cathepsin D* did not correlate with relative activity of cathepsin D, even if relative expression of *cathepsin L* was correlated with relative activity of cathepsin L during spawning and post-spawning of rainbow trout. In studies II and III, an increase in *cathepsin B* and/or *L* relative gene expression(s) was always positively correlated with muscle degradation measured as percentage of myofibre-myofibre detachments, and/or fillet softness. These results reflect the complexity of the system and the multiplicity of the control points during the post-translational sorting of newly synthesized cathepsins to endosomes/lysosomes (Béchet et al., 2005).

In study II, very high levels of dietary n-3 fatty acids influenced total activity of both cathepsins B and L in the muscle, but as for relative gene expression, only *cathepsin B* was affected. In study III, pre-slaughter crowding stress had an impact on cathepsin B total activity, but not cathepsin L activity, but *cathepsin L* relative gene expression was influenced. The complexity of the cathepsin system thus appeared during all three studies, as from our results, it seemed that cathepsin B and cathepsin L did not react the same way to the different factors to which salmon were subjected.

With our three studies presented in this chapter, it was difficult to point out the direct factors determining fillet texture. In study I, fillet texture was not influenced by the super-chilling treatment, even though muscle integrity (myofibre-myofibre detachments) was significantly affected. This could be due to the fact that fillet texture was measured 168 h *post-mortem*, and that at this stage, super-chilling did not have any impact on fillet quality anymore, but other parameters could have taken over the impact on the fillets. Indeed, in study III, pre-slaughter long-term crowding stress accelerated fillet softening in the first phases of chilled storage, but after 24 h *post-mortem*, no significant impact of stress on fillet texture was noticed anymore.

In study III, a significant positive correlation was found between fillet softness and cathepsin B activity, and *cathepsin L* relative gene expression. The amount of collagen and cross-linking in the muscle has shown to influence fish fillet texture (Ando et al., 1991, 1992a and 1992b; Johnston et al., 2006; Suarez et al., 2003) and cathepsins, especially cathepsin L, can degrade collagen (Yamashita and Konagaya, 1991). According to our results and in accordance to Godiksen et al. (2009), cathepsins could thus play an important role in fillet firmness.

Studies I and III indicated that fillet texture could be influenced by different synergic factors, among them the level of cathepsin B and/or L activity, *cathepsin L* gene expression, myofibre-myofibre and myofibre-myocommata detachments, with significant correlations between these traits and fillet firmness in study III. In study I, 168 h *post-mortem*, there were no differences in fillet firmness between super-chilled and control fillets, although a lower level of cathepsin B + L activity in the lysosomal fraction (pellet), and a higher percentage of myofibre-myofibre detachments were found in the super-chilled fillets. The percentage of myofibre-myocommata detachments was also the same as in the control fillets. A decrease in fillet firmness in the super-chilled fillets was thus expected, but it was not the case. Skjervold et al. (2001a) and Taylor et al. (2002) already mentioned the complexity of the mechanisms behind fish fillet texture and the fact that fish fillet texture could result from different factors; our results confirming this observation.

3.10. Other Considerations on Atlantic Salmon Flesh Quality

In our three studies (I, II and II), we did not measure calpain activity or their inhibitor, but their role in salmon muscle degradation, especially in the first phase *post-mortem*, cannot be totally excluded. This hypothesis agrees with Delbarre-Ladrat et al. (2004b) and Goll et al., (1992), who also pointed out that calpains could be involved in the first degradation of the myofibrillar proteins, followed by the action of cathepsins. Yamashita and Konagaya (1991) indeed noticed that cathepsin B could degrade denatured proteins very well. It is not excluded that other proteases like collagenases could also be involved in fillet softness, as collagen and collagen cross-linking species have been shown to influence fish fillet texture (Ando et al., 1991, 1992a and 1992b; Espe et al., 2004; Hultmann and Rustad, 2007; Johnston et al., 2006).

In general, the significant effects of the different factors studied here (super-chilling, dietary lipids and pre-slaughter stress) on Atlantic salmon quality were mainly noticed in the first phases of chilled storage, until around 24 h - 48 h *post-mortem* and usually, after 1 week of storage, significant differences between control and treated salmon disappeared, which is in accordance with other authors' suggestions (Erikson et al., 1997a; Skjervold et al., 2001a; Stien et al., 2005). This could be due to the fact that after *rigor*, biochemical factors do not have the same power on muscle integrity and that their impact is minimal in front of other factors like bacterial spoilage or mechanical degradation due to *rigor* stiffness only. These results thus have their importance especially when dealing with *pre-rigor* fillet quality.

The percentage of myofibre-myocommata detachments was not significantly influenced by super-chilling and dietary lipids during storage time until at least up to 4 weeks *post-mortem* (study I and II). Taylor et al. (2002) suggested that myofibre-myocommata detachments were associated with *rigor* stiffness, while myofibre-myofibre detachments were more related to muscle texture. This observation could explain why some previous studies also did not show any changes in myofibre-myocommata attachments before 5 and 7 days in Atlantic salmon and wolfish respectively (Ofstad et al., 2006; Taylor et al., 2002). In study III, pre-slaughter crowding stress (SS and LS, but especially SS) however had a significant impact during the storage period until 192 h *post-mortem*, increasing the percentage of myofibre-myocommata detachments. The difference between the results obtained on studies I, and II, and III could be due to muscle pH, particularly low in the stressed salmon, but probably higher in the super-chilled and different lipid fed salmon. Lavéty et al. (1988) showed indeed a strong negative correlation between muscle pH and degree of gaping in salmon and rainbow trout, and the lower the pH, the easier to break the myocommata. Low muscle pH could have lead to shrinkage of the myofibres, increasing the tensions in the connective tissue and resulting in a higher percentage of myofibre-myocommata detachments.

Conclusion

Cathepsin B and L activities and/or gene expressions were influenced by the main factors tested in the three studies of this PhD project (super-chilling, dietary fatty acids and pre-slaughter stress). Our results indicated that formation of ice crystals, oxidative stress, and muscle pH after stress can apparently play a role in the integrity of the lysosomal system and the action of the lysosomal catheptic proteases on muscle structure, both at the myofibrillar and connective tissue levels.

Cathepsin B and L activities and/or gene expressions were in general significantly negatively correlated to muscle pH, percentage of myofibre-myofibre and myofibre-myocommata attachments and fillet firmness, suggesting their potential role in the determination of Atlantic salmon muscle integrity and flesh quality. The possible action of cathepsins B and/or L in relation to muscle degradation seemed to concern both myofibrillar and connective tissue proteins. Cathepsin B and L are known to be able to degrade myofibrillar proteins. According to our results, cathepsin B and L activities and/or gene expressions were also strongly correlated to the degradation of the extra cellular matrix (myofibre-myofibre and myofibre-myocommata detachments). Denaturation of proteoglycans and other glycoproteins, which are components of the connective tissue, might therefore be involved.

Cathepsin B and cathepsin L were not influenced the same way by the different factors studied here, and cathepsin activity was not always correlated with *cathepsin* gene expression. This shows the complexity of the lysosomal system and the mechanisms behind the action of lysosomal cathepsins in Atlantic salmon flesh.

REFERENCES

Alami-Durante, H. & Rescan, P. Y. (2003). Typologie et ontogenèse des fibres musculaires chez les poissons. *INRA Productions Animales, 16*, 145-155. *http://www.inra.fr /productions-animales/ an2003/ num232/ alami/ ha232. htm - Last access 16.11.2006.*

Alizadeh, P., Smit-McBride, Z., Oltjen, S. L. & Hjelmeland, L. M. (2006). Regulation of cysteine cathepsin expression by oxidative stress in the retinal pigment epithelium/choroid of the mouse. *Experimental Eye Research, 83*, 679-687.

Allison, A. C. (1975). Lysosomes in Biology and Pathology, Vol.2. *North Holland Elsevier*, Amsterdam, New-York, 178-204.

American Society for Quality Source. (2009). Quality. *http://www.asq.org/* glossary/q.html - Last access 06.04.2009.

Ando, M., Toyohara, H., Shimizu, Y. & Sakaguchi, M. (1991). Postmortem Tenderization of Rainbow-Trout (Oncorhynchus-Mykiss) Muscle Caused by Gradual Disintegration of the Extracellular-Matrix Structure. *J. Sci. Food Agric. 55*, 589-597.

Ando, M., Toyohara, H. & Sakaguchi, M. (1992a). *Three-dimensional structure of collagen fibrillar network of pericellular connective-tissue in association with firmness of fish muscle Nippon Suisan Gakkaishi, 58*, 1361-1364.

Ando, M., Toyohara, H. & Sakaguchi, M. (1992b). Postmortem tenderization of Rainbow-Trout muscle caused by the desintegration of collagen-fibers in the pericellular connective-tissue. *Nippon Suisan Gakkaishi, 58*, 567-570.

Ando, M., Joka, M., Mochizuki, S., Satoh, K. I., Tsukamasa, Y. & Makinodan, Y. (2001). Influence of death struggle on the structural changes in chub mackerel muscle during chilled storage. *Fish. Sci., 67*, 744-751.

Aoki, T., Yamashita, T. & Ueno, R. (1995). Purification and Characterization of a Cathepsin B-Like Enzyme from Mackerel White Muscle. *Fish. Sci., 61*, 121-126.

Aoki, T. & Ueno, R. (1997). Involvement of cathepsins B and L in the post-mortem autolysis of mackerel muscle. *Food Research International, 30*, 585-591.

Bagni, M., Civitareale, C., Priori, A., Ballerini, A., Finola, M., Brambilla, G. & Marino, G. (2007). Pre-slaughter crowding stress and killing procedures affecting quality and welfare in sea bass (Dicentrarchus labrax) and sea bream (Sparus aurata). *Aquaculture, 263*, 52-60.

Bahuaud, D., Mørkøre, T., Langsrud, Ø., Sinnes, K., Veiseth, E., Ofstad, R. & Thomassen, M. S. (2008). Effects of -1.5 °C Super-chilling on quality of Atlantic salmon (*Salmo salar) pre-rigor* Fillets: Cathepsin activity, muscle histology, texture and liquid leakage. *Food Chemistry, 111*, 329-339.

Bahuaud, D., Mørkøre, T., Østbye, T. K., Veiseth-Kent, E., Thomassen, M. S. & Ofstad, R. (2009a). Muscle structure responses and lysosomal cathepsins B and L in farmed Atlantic salmon (Salmo salar L.) pre- and post-rigor fillets exposed to short and long-term crowding stress. *Food Chemistry, 118*, 602-615.

Bahuaud, D., Østbye, T. K., Torstensen, B. E., Rørå, M. B., Ofstad, R., Veiseth, E., Thomassen, M. S. & Ruyter, B. (2009b). Atlantic salmon (Salmo salar) muscle structure integrity and lysosomal cathepsins B and L influenced by dietary n-6 and n-3 fatty acids. *Food Chemistry, 114*, 1421-1432.

Barrett, A. J. & Kirschke, H. (1981). Cathepsin-B, cathepsin-H, and cathepsin-L. *Methods in Enzymology*, *80*, 535-561.

Béchet, D., Tassa, A., Taillandier, D., Combaret, L. & Attaix, D. (2005). Lysosomal proteolysis in skeletal muscle. *The International Journal of Biochemistry & Cell Biology*, *37*, 2098-2114.

Bell, J. G., McEvoy, J., Tocher, D. R., McGhee, F., Campbell, P. J. & Sargent, J. R. (2001). Replacement of fish oil with rapeseed oil in diets of Atlantic salmon (Salmo salar) affects tissue lipid compositions and hepatocyte fatty acid metabolism. *Journal of Nutrition*, *131*, 1535-1543.

Bell, J. G., Henderson, R. J., Tocher, D. R., McGhee, F., Dick, J. R., Porter, A., Smullen, R. P. & Sargent, J. R. (2002). Substituting fish oil with crude palm oil in the diet of Atlantic salmon (Salmo salar) affects muscle fatty acid composition and hepatic fatty acid metabolism. *Journal of Nutrition*, *132*, 222-230.

Bell, J. G., Tocher, D. R., Henderson, R. J., Dick, J. R. & Crampton, V. O. (2003). Altered fatty acid compositions in Atlantic salmon (Salmo salar) fed diets containing linseed and rapeseed oils can be partially restored by a subsequent fish oil finishing diet. *Journal of Nutrition*, *133*, 2793-2801.

Bello, R. A., Luft, J. H. & Pigott, G. M. (1982). Ultrastructural-Study of Skeletal Fish Muscle after Freezing at Different Rates. *Journal of Food Science*, *47*, 1389-1394.

Brodtkorb, B. T., Rosenlund, G. & Lie, Ø. (1997). Effects of 20:5n-3 and 22:6n-3 on tissue lipid composition in juvenile Atlantic salmon, Salmo salar, – with emphasis on brain and eye. *Aquaculture Nutrition*, *3*, 175-187.

Brownell, A. K. W. & Engel, A. G. (1978). Experimental lipid storage myopathy : A quantitative ultrastructural and biochemical study. *Journal of the Neurological Sciences*, *35*, 31-41.

Bøknæs, N., Osterberg, C., Nielsen, J. & Dalgaard, P. (2000). Influence of freshness and frozen storage temperature on quality of thawed cod fillets stored in modified atmosphere packaging. *Lebensmittel-Wissenschaft Und-Technologie-Food Science and Technology*, *33*, 244-248.

Caballero, M. J., Obach, A., Rosenlund, G., Montero, D., Gisvold, M. & Izquierdo, M. S. (2002). Impact of different dietary lipid sources on growth, lipid digestibility, tissue fatty acid composition and histology of rainbow trout, Oncorhynchus mykiss. *Aquaculture*, *214*, 253-271.

Chéret, R., Hernandez-Andres, A., Delbarre-Ladrat, C., de Lamballerie, M. & Verrez-Bagnis, V. (2006). Proteins and proteolytic activity changes during refrigerated storage in sea bass (Dicentrarchus labrax L.) muscle after high-pressure treatment. *European Food Research and Technology*, *222*, 527-535.

Chéret, R., Delbarre-Ladrat, C., de Lamballerie-Anton, M. & Verrez-Bagnis, V. (2007). Calpain and cathepsin activities in post mortem fish and meat muscles. *Food Chemistry*, *101*, 1474-1479.

Choi, K. H. & Park, C. S. (1991). Changes in the Ca^{2+} and Mg^{2+}-dependent adenosine triphosphate activity and ultrastructure of marine fishes by partial freezing III. Changes in the ultrastructure of muscle tissues of yellowtail during low temperature preservation. *Journal of Korean Society of Food and Nutrition*, *20*, 629-638.

Delbarre-Ladrat, C., Verrez-Bagnis, V., Noël, J. & Fleurence, J. (2004a). Proteolytic potential in white muscle of sea bass (Dicentrarchus labrax L.) during post mortem storage on ice:

time-dependent changes in the activity of the components of the calpain system. *Food Chemistry*, *84*, 441-446.

Delbarre-Ladrat, C., Verrez-Bagnis, V., Noel, J. & Fleurence, J. (2004b). Relative contribution of calpain and cathepsins to protein degradation in muscle of sea bass (Dicentrarchus labrax L.). *Food Chemistry*, *88*, 389-395.

Delbarre-Ladrat, C., Chéret, R., Taylor, R. & Verrez-Bagnis, V. (2006). Trends in postmortem aging in fish: Understanding of proteolysis and disorganization of the myofibrillar structure. *Critical Reviews in Food Science and Nutrition*, *46*, 409-421.

Dion, P. (1997). In Pursuit of Prime Quality Seafood - Introduction to Super-chilling, pp. *http://members.aol.com/pdionpda/sc_intro.htm* (last access 05.09.2007).

Duun, A. S. & Rustad, T. (2007). Quality changes during superchilled storage of cod (Gadus morhua) fillets. *Food Chemistry*, *105*, 1067-1075.

Duun, A. S. & Rustad, T. (2008). Quality of superchilled vacuum packed Atlantic salmon (Salmo salar) fillets stored at -1.4 and -3.6 [degree sign]C. *Food Chemistry*, *106*, 122-131.

Einen, O. & Thomassen, M. S. (1998). Starvation prior to slaughter in Atlantic salmon (Salmo salar) II. White muscle composition and evaluation of freshness, *texture and colour characteristics in raw and cooked fillets. Aquaculture*, *169*, 37-53.

Einen, O., Guerin, T., Fjæra, S. O. & Skjervold, P. O. (2002a). Freezing of pre-rigor fillets of Atlantic salmon. *Aquaculture*, *212*, 129-140.

Einen, O., Tomter, R. & Fjæra, S. O. (2002b). Contraction and quality changes of pre-rigor fillets of Atlantic salmon as affected by dietary oil and live-chilling, *International Symposium on Nutrition and Feeding of Fish, Rhodes*, 2-7 June 2002.

Eksportutvalget for fisk, (2009). Verden og Vi, Tall og Fakta 2008. http://www.seafood.no/_public/Seafood.no/verdenogvi/full.html - Last access 29.04.2009.

Erikson, U., Beyer, A. R. & Sigholt, T. (1997a). Muscle high-energy phosphates and stress affect K-values during ice storage of Atlantic salmon (Salmo salar). *Journal of Food Science*, *62*, 43-47.

Erikson, U., Sigholt, T. & Seland, A. (1997b). Handling stress and water quality during live transportation and slaughter of Atlantic salmon (Salmo salar). *Aquaculture*, *149*, 243-252.

Erikson, U., Sigholt, T., Rustad, T., Einarsdottir, I. E. & Jørgensen, L. (1999). Contribution of bleeding to total handling stress during slaughter of Atlantic salmon. *Aquaculture International*, *7*, 101-115.

Erikson, U. & Misimi, E. (2008). Atlantic salmon skin and fillet color changes effected by perimortem handling stress, rigor mortis, and ice storage. *Journal of Food Science*, *73*, C50-C59.

Espe, M., Ruohonen, K., Bjornevik, M., Froyland, L., Nortvedt, R. & Kiessling, A. (2004). Interactions between ice storage time, collagen composition, gaping and textural properties in fanned salmon muscle harvested at different times of the year. *Aquaculture*, *240*, 489-504.

FAO, (2005). Postmortem changes in fish. In Quality and quality changes in fresh fish, Chapter 5. http://www.fao.org//docrep/V7180E/v7180e06.htm - Last access 21.09.2005

Fik, M., Surowka, K. & Leszczynskafik, A. (1988). *Application of Superchilling to Prolong the Keeping Time of Rainbow-Trout. Nahrung-Food*, *32*, 291-300.

Gallart-Jornet, L., Rustad, T., Barat, J. M., Fito, P. & Escriche, I. (2007). Effect of superchilled storage on the freshness and salting behaviour of Atlantic salmon (Salmo salar) fillets. *Food Chemistry*, *103*, 1268-1281.

García-Buñuel, L. (1984). Lipid peroxidation in alcoholic myopathy and cardiomyopathy. *Medical Hypotheses*, *13*, 217-231.

Gil, M., Guerrero, L. & Sarraga, C. (1999). The effect of meat quality, salt and ageing time on biochemical parameters of dry-cured Longissimus dorsi muscle. *Meat Science*, *51*, 329-337.

Gjedrem, T. (2000). Genetic improvement of cold-water fish species. *Aquaculture Research*, *31*, 25-33.

Godiksen, H., Morzel, M., Hyldig, G. & Jessen, F. (2009). Contribution of cathepsins B, L and D to muscle protein profiles correlated with texture in rainbow trout (Oncorhynchus mykiss). *Food Chemistry*, *113*, 889-896.

Goll, D. E., Thompson, V. F., Taylor, R. G. & Christiansen, J. A. (1992). *Role of the calpain system in muscle growth*, 225-237.

Goll, D. E., Thompson, V. F., Li, H. Q., Wei, W. & Cong, J. Y. (2003). The calpain system. *Physiological Reviews*, *83*, 731-801.

Gomez, T. M. & Calvelo, A. (1982). On the ice growth mechanisms during beef freezing, *Proceedings of the Meeting of Commissions C2, D1, D2 and D3 of the IIF*, Hamilton, New Zealand, 83-91.

Gomez-Guillen, M. C., Montero, P., Hurtado, O. & Borderias, A. J. (2000). Biological characteristics affect the quality of farmed Atlantic salmon and smoked muscle. *Journal of Food Science*, *65*, 53-60.

Guillerm-Regost, C., Haugen, T., Nortvedt, R., Carlehog, M., Lunestad, B. T., Kiessling, A. & Rørå, A. M. B. (2006). Quality characterization of farmed Atlantic halibut during ice storage. *Journal of Food Science*, *71*, S83-S90.

Hansen, A. Å., Mørkøre, T., Rudi, K., Olsen, E. & Eie, T. (2007). Quality changes during refrigerated storage of MA-packaged pre-rigor fillets of farmed Atlantic cod (Gadus morhua L.) using traditional MAP, CO2 emitter, and vacuum. *Journal of Food Science*, *72*, M423-M430.

Hansen, A. Å. (2008). Reduced headspace volume of modified atmosphere packaged fresh salmon (*Salmo salar* L.) and cod (*Gadus morhua* L.) by use of a carbon dioxide emitter. PhD Thesis 2008:48. *Norwegian University of Life Sciences - Department of Animal and Aquacultural Sciences*. Ås, Norway.

Haugland, A. (2001). Super Chilling of Fresh Marine Products. Glimt Newsletter - SINTEF Energy Research Nr. 2-2001 Refrigeration and Conditioning, http://www.energy.sintef.no /publ/Nyhetsbrev/2001/Klima-eng%2202.pdf - *Last access 07.04.2009*

Haugland, A., Hardarson, V., Hemmingsen, A. K. T., Magnussen, O. M., Duun, A. S. & Rustad, T. (2006). Advances in Superchilling - Process Characteristics and Product Quality, Sintef Energy Research, CEFood 2006, Sofia, http:// www. sintef. no/ upload/ Energiforskning/ Energiprosesser/ Superkj %C3%B8ling/ Superchilling_ CEFood_ 2006_ final.pdf - *Last access 04.09.2007.*

Headlam, H. A., Gracanin, M., Rodgers, K. J. & Davies, M. J. (2006). Inhibition of cathepsins and related proteases by amino acid, peptide, and protein hydroperoxides. *Free Radical Biology and Medicine*, *40*, 1539-1548.

Hemmingsen, A. K. T. & Nordtvedt, T. S. (2007). Pilot study - Competitive Food Processing in norway (TR A6480) - Superchilling of pork and salmon fillets - a feasability study. *http://www.energy.sintef.no/publ/* rapport/ 07/tr6480.htm - Last access 19.06.2007.

Hultmann, L. (2003). PhD thesis - Endogenous proteolytic enzymes - Studies of their impact on fish muscle proteins and texture, Department of Biotechnology. *Faculty of Natural Sciences and Technology*, Trondheim, Norway.

Hultmann, L. Rustad, T. (2004). Iced storage of Atlantic salmon (Salmo salar) - effects on endogenous enzymes and their impact on muscle proteins and texture. *Food Chemistry*, *87*, 31-41.

Hultmann, L. & Rustad, T. (2007). Effects of temperature abuse on textural properties and proteolytic activities during post mortem iced storage of farmed Atlantic cod (Gadus morhua). *Food Chemistry*, *104*, 1687-1697.

Huss, H. (1988). Fresh fish quality and quality changes. H. Huss & *Technological Laboratory Ministry of Agriculture and Fisheries,* Fisheries No. 29, FAO.

Inteaz, A. (2000). Food Quality Assurance. ISO 9000:2000. CRC press http://books.google.no/books?id=pk6Cjm5cYyMC&pg=PA20&lpg=PA20&dq=ISO+9000+the+degree+to+which+a+set+of+inherent+characteristic+fulfills+requirements&source=bl&ots=LwhQUB5qit&sig=rLL4_5Bcff3xHCs6yHoxBBNrgU4&hl=no&ei=uFgVSs6jOMqS_Qalv6ndDA&sa=X&oi=book_result&ct= result&resnum=1#PPP1,M1, *Last access 21.05.2009.*

Izquierdo, M. S., Montero, D., Robaina, L., Caballero, M. J., Rosenlund, G. & Gines, R. (2005). Alterations in fillet fatty acid profile and flesh quality in gilthead seabream (Sparus aurata) fed vegetable oils for a long term period. Recovery of fatty acid profiles by fish oil feeding. *Aquaculture*, *250*, 431-444.

Jensen, M. H. (2007). Effects of super-chilling on fresh fish storage quality, Aker Seafoods, European Fish and Seafoods Conference, Campden and *Chorleywood Food Research Association Group ICO Conference Centre*, London.

Jiang, S. T., Wang, Y. T. & Chen, C. S. (1992). Lysosomal-Enzyme Effects on the Postmortem Changes in Tilapia (Tilapia-Nilotica X T-Aurea) Muscle Myofibrils. *Journal of Food Science*, *57*, 277-&.

Johnston, I. A. (1999). Muscle development and growth: potential implications for flesh quality in fish. *Aquaculture*, *177*, 99-115.

Johnston, I. A., Alderson, R., Sandham, C., Dingwall, A., Mitchell, D., Selkirk, C., Nickell, D., Baker, R., Robertson, B., Whyte, D. & Springate, J. (2000). Muscle fibre density in relation to the colour and texture of smoked Atlantic salmon (Salmo salar L.). *Aquaculture*, *189*, 335-349.

Johnston, I. A., Li, X., Vieira, V. L. A., Nickell, D., Dingwall, A., Alderson, R., Campbell, P. & Bickerdike, R. (2006). Muscle and flesh quality traits in wild and farmed Atlantic salmon. *Aquaculture*, *256*, 323-336.

Johnston, I. A., Bickerdike, R., Li, X., Dingwall, A., Nickell, D., Alderson, R. & Campbell, P. (2007). Fast growth was not associated with an increased incidence of soft flesh and gaping in two strains of Atlantic salmon (Salmo salar) grown under different environmental conditions. *Aquaculture*, *265*, 148-155.

Junqueira L. C., Carneiro J. & Kelley R. O. (1989a). Connective tissue. Chapter 5, In: Basic Histology, Sixth Edition. *Appleton & Lange*, 91-121.

Junqueira L. C., Carneiro J. & Kelley R. O. (1989b). Muscle Tissue. In: Basic Histology, Sixth Edition. *Appleton & Lange*, 194-213.

Kharbanda, K. K., McVicker, D. L., Zetterman, R. K. & Donohue, T. M. (1995). Ethanol consumption reduces the proteolytic capacity and protease activities of hepatic lysosomes. Biochimica et Biophysica Acta (BBA) - *General Subjects*, *1245*, 421-429.

Kharbanda, K. K., McVicker, D. L., Zetterman, R. K. & Donohue, T. M. (1996). Ethanol consumption alters trafficking of lysosomal enzymes and affects the processing of procathepsin L in rat liver. *Biochimica et Biophysica Acta (BBA) - General Subjects*, *1291*, 45-52.

Kiessling, A., Espe, M., Ruohonen, K. & Mørkøre, T. (2004). Texture, gaping and colour of fresh and frozen Atlantic salmon flesh as affected by pre-slaughter iso-eugenol or CO2 anaesthesia. *Aquaculture*, *236*, 645-657.

Kiessling, A., Stien, L. H., Torslett, V., Suontarna, J. & Slinde, E. (2006). Effect of pre- and post-mortem temperature on rigor in Atlantic salmon muscle as measured by four different techniques. *Aquaculture*, *259*, 390-402.

Kiessling, A., Bjørnevik, M., Thomassen, M., Rørå, M. B., Mørkøre, T., Roth, B., Erikson, U. & Jordheim, O. (2007). From Cage to Table. In: Aquaculture: *from cage to consumption The research council of Norway, Oslo*, Norway, 45-63.

Kim, S. K., Park, P. J., Kim, J. B. & Shahidi, F. (2002). Purification and characterization of a collagenolytic protease from the filefish. Novoden modestrus. *Journal of Biochemistry and Molecular Biology*, *335(2)*, 165-171.

Kirkland, R. A., Adibhatla, R. M., Hatcher, J. F. & Franklin, J. L. (2002). Loss of cardiolipin and mitochondria during programmed neuronal death: Evidence of a role for lipid peroxidation and autophagy. *Neuroscience*, *115(2)*, 587-602.

Kirschke, H. (1981). On the substrate-specificity of cathepsin L. *Acta Biologica Et Medica Germanica*, *40*, 1427-1431.

Kirschke, H., Wood, L., Roisen, F. J. & Bird, J. W. C. (1983). Activity of Lysosomal Cysteine Proteinase During Differentiation of Rat Skeletal-Muscle. *Biochemical Journal*, *214*, 871-877.

Kjær, M. A., Vegusdal, A., Gjøen, T., Rustan, A. C., Todorevic, M. & Ruyter, B. (2008). Effect of rapeseed oil and dietary n-3 fatty acids on triacylglycerol synthesis and secretion in Atlantic salmon hepatocytes. *Biochimica Et Biophysica Acta-Molecular and Cell Biology of Lipids*, *178*, *1*, 112-122.

Koohmaraie, M. (1992). The role of Ca2+-dependent proteases (calpains) in post mortem proteolysis and meat tenderness. *Biochimie*, *74*, 239-245.

Koohmaraie, M. (1996). Biochemical factors regulating the toughening and tenderization processes of meat. *Meat Science*, *43*, 193-201.

Kurz, T., Terman, A., Gustafsson, B. & Brunk, U. T. (2008). Lysosomes and oxidative stress in aging and apoptosis. Biochimica et Biophysica Acta (BBA) – *General Subjects*, *1780*, 1291-1303.

Köhler, A. (1989). Cellular effects of environmental contamination in fish from the River Elbe and the North Sea. *Marine Environmental Research*, *28*, 417-424.

Köhler, A. (1991). Lysosomal perturbations in fish liver as indicators for toxic effects of environmental pollution. Comparative Biochemistry and Physiology Part C: *Comparative Pharmacology*, *100*, 123-127.

Ladrat, C., Verrez-Bagnis, V., Noel, J. & Fleurence, J. (2003). In vitro proteolysis of myofibrillar and sarcoplasmic proteins of white muscle of sea bass (Dicentrarchus labrax L.): effects of cathepsins B, D and L. *Food Chemistry*, *81*, 517-525.

Langsrud, Ø., Jørgensen, K., Ofstad, R. & N_s, T. (2007). Analyzing designed experiments with multiple responses. *Journal of Applied Statistics*, *34*, 1275-1296.

Lavéty, J., Afolabi, O. A. & Love, R. M. (1988). The connective tissues of fish . IX. *Gaping in farmed species International Journal of Food Science and Technology*, *23*, 23-30.

Lee, C. M. & Toledo, R. T. (1984). Comparison of Shelf-Life and Quality of Mullet Stored at Zero and Subzero Temperature. *Journal of Food Science*, *49*, 317-322.

Losada, V., Piñeiro, C., Barros-Velazquez, J. & Aubourg, S. P. (2005). Inhibition of chemical changes related to freshness loss during storage of horse mackerel (Trachurus trachurus) in slurry ice. *Food Chemistry*, *93*, 619-625.

Lowe, T. E., Ryder, J. M., Carragher, J. F. & Wells, R. M. G. (1993). Flesh quality in snapper, *Pagrus-Auratus*, affected by capture stress. *Journal of Food Science*, *58*, 770-&.

Macdonald, R. G. & Cidlowski, J. A. (1982). Glucocorticoids inhibit precursor incorporation into protein in splenic lymphocytes by stimulating proteindegradation and expanding intracellular amino-acid pools. *Biochimica Et Biophysica Acta*, *717*, 236-247.

Mackie, I. M. (1993). The effect of freezing on flesh proteins. *Food Review International*, *9*, 575-610.

Martinsen, T. S. (2006). Impact of handling stress on contraction and post-mortem metabolism of pre-rigor filleted Atlantic salmon (Salmo salar L.). Master Thesis, Department of Animal and Aquacultural Sciences, *Norwegian University of Life Sciences*, Ås, 49.

Martino, M. N. & Zaritzky, N. E. (1988). Ice crystal size modifications during frozen beef storage. *Journal of Food Science*, *53*, 1631-1637.

Martino, M. N., Otero, L., Sanz, P. D. & Zaritzky, N. E. (1998). Size and location of ice crystals in pork frozen by high-pressure-assisted freezing as compared to classical methods. *Meat Science*, *50*, 303-313.

Misimi, E., Erikson, U., Digre, H., Skavhaug, A. & Mathiassen, J. R. (2008). Computer vision-based evaluation of pre- and postrigor changes in size and shape of Atlantic cod (Gadus morhua) and Atlantic salmon (Salmo salar) fillets during rigor mortis and ice storage: Effects of perimortem handling stress. *Journal of Food Science*, *73*, E57-E68.

Mitchie, I. (2001). Causes of downgrading in the salmon farming industry. Farmed fish quality. Kestin SC., Warris PD, editors, Oxford Fishing New Books.

Molina-Garcia, A. D., Otero, L., Martino, M. N., Zaritzky, N. E., Arabas, J., Szczepek, J. & Sanz, P. D. (2004). Ice VI freezing of meat: supercooling and ultrastructural studies. *Meat Science*, *66*, 709-718.

Montero, D., Robaina, L., Caballero, M. J., Gines, R. & Izquierdo, M. S. (2005). Growth, feed utilization and flesh quality of European sea bass (Dicentrarchus labrax) fed diets containing vegetable oils: A time-course study on the effect of a re-feeding period with a 100% fish oil diet. *Aquaculture*, *248*, 121-134.

Moore, M. N. & Lowe, D. M. (1985). Effects of Stress and Pollution on Marine Animals. *Praeger Publishers*, New-York, 190-204.

Moore, M. N. (1988). Mar. Ecol. *Progr. Ser*, *46*, 81-89.

Moore, M. N., Icarus Allen, J. & McVeigh, A. (2006). Environmental prognostics: An integrated model supporting lysosomal stress responses as predictive biomarkers of animal health status. *Marine Environmental Research*, *61*, 278-304.

Morzel, M., Sohier, D. & Van de Vis, H. (2003). Evaluation of slaughtering methods for turbot with respect to animal welfare and flesh quality. *J. Sci. Food Agric*, *83*, 19-28.

Mundheim, H., Aksnes, A. & Hope, B. (2004). Growth, feed efficiency and digestibility in salmon (Salmo salar L.) fed different dietary proportions of vegetable protein sources in combination with two fish meal qualities. *Aquaculture*, *237*, 315-331.

Mørkøre, T. & Rørvik, K. A. (2001). Seasonal variations in growth, feed utilisation and product quality of farmed Atlantic salmon (Salmo salar) transferred to seawater as 0+smolts or 1+smolts. *Aquaculture*, *199*, 145-157.

Mørkøre, T. (2006). Relevance of dietary oil source for contraction and quality of pre-rigor filleted Atlantic cod, Gadus morhua. *Aquaculture*, *251*, 56-65.

Mørkøre, T., Bahuaud, D., Bjerke, M., Eie, T., Gullsvåg, P. E. & Hansen, A. Å., et al., (2007). Quality of Pre-rigor Filleted and Superchilled Atlantic Salmon (Salmo salar). In 37th WEFTA annual meeting, seafood: Source of health and consumer well-being, Lisbon, Portugal.

Mørkøre, T. (2008). Tekstur i oppdrettslaks - Kunnskapsstatus og forhold som bidrar til fastere fillet. Norske Sjømatbedrifters Landsforening. *www.nsl.no/filer/fou/loe/loe5.pdf* - Last access 01.04.2009.

Mørkøre, T., Mazo, T, P. I., Tahirovic, V. & Einen, O. (2008). Impact of starvation and handling stress on rigor development and quality of Atlantic salmon (Salmon salar L). *Aquaculture*, *277*, 231-238.

Nakayama, T., Toyoda, T. & Ooi, A. (1994). Physical property of carp muscle during rigor tension generation. *Fisheries Science*, *60*, 717-721.

Nakayama, T., Goto, E. & Ooi, A. (1996). Observation of characteristic muscle structure related to delay in red sea-bream rigor mortis by spinal cord destruction. *Fish. Sci.*, *62*, 977-984.

Nanton, D. A., Vegusdal, A., Rørå, A. M. B., Ruyter, B., B_verfjord, G. & Torstensen, B. E. (2007). Muscle lipid storage pattern, composition, and adipocyte distribution in different parts of Atlantic salmon (Salmo salar) fed fish oil and vegetable oil. *Aquaculture*, *265*, 230-243.

Norconserv. *Forskningsrådet*, Nytt fra HAVBRUK nr. 3. Nov. 2007.

Nortvedt, R., Espe, M., Gribbestad, I. S., Jørgensen, L., Karlsen, Ø., Otterå, H., Rørå, M. B., Stien, L. H. & Sørensen, N. K. (2007). High-quality Seafood Products based on Ethical and Sustainable Production. In: Aquaculture Research: from cage to consumption. *The research council of Norway*, Oslo, Norway, 28-44.

Nowlan, S. S., Dyer, W. J. & Keith, R. A. (1974). Superchilling - New Application for Preserving Freshness of Fish Filets During Marketing. *Canadian Institute of Food Science and Technology Journal-Journal De L Institut Canadien De Science Et Technologie Alimentaires*, *7*, A16-A19.

Ofstad, R., Olsen, R. L., Taylor, R. & Hannesson, K. O. (2006). Breakdown of intramuscular connective tissue in cod (Gadus morhua L.) and spotted wolffish (Anarhichas minor O.) *related to gaping. Lwt-Food Science and Technology*, *39*, 1143-1154.

Okamoto, A., Hamada, Y., Miura, K., Nonaka, T., Kuwahara, K., Ohsako, K., Mishima, T. & Tachibana, K. (2006). Influence of killing procedures and storage temp eratures on post-mortem changes in the muscle of cultured three-line grunt Parapristipoma trilineatum. *Nippon Suisan Gakkaishi*, *72*, 918-923.

Okitani, A., Matsukura, U., Kato, H. & Fujimaki, M. (1980). Purification and Some Properties of a Myofibrillar Protein-Degrading Protease, Cathepsin-L, from Rabbit Skeletal-Muscle. *Journal of Biochemistry*, *87*, 1133-1143.

Olsen, R. E. & Henderson, R. J. (1997). Muscle fatty acid composition and oxidative stress indices of Arctic charr, Salvelinus alpinus (L.), in relation to dietary polyunsaturated fatty acid levels and temperature. *Aquaculture Nutrition*, *3*, 227-238.

Olsen, R. E., Suontama, J., Langmyhr, E., Mundheim, H., Ringo, E., Melle, W., Malde, M. K. & Hemre, G. I. (2006). The replacement of fish meal with Antarctic krill, Euphausia superba in diets for Atlantic salmon, Salmo salar. *Aquaculture Nutrition*, *12*, 280-290.

Papa, I., Taylor, R. G., Astier, C., Ventre, F., Lebart, M. C., Roustan, C., Ouali, A. & Benyamin, Y. (1997). Dystrophin cleavage and sarcolemma detachment are early post mortem changes on pass (Dicentrarchus labrax) white muscle. *Journal of Food Science*, *62*, 917-921.

Petrosillo, G., Ruggiero, F. M., Pistolese, M. & Paradies, G. (2001). Reactive oxygen species generated from the mitochondrial electron transport chain induce cytochrome c dissociation from beef-heart submitochondrial particles via cardiolipin peroxidation. Possible role in the apoptosis. *FEBS Letters*, *509*, 435-438.

Poli, B. M., Parisi, G., Scappini, F. & Zampacavallo, G. (2005). Fish welfare and quality as affected by pre-slaughter and slaughter management. *Aquaculture International*, *13*, 29-49.

Pote, M. S. & Altekar, W. (1980). Increase in rat serum aldolase levels induced by x-irradiation. *Indian Journal of Biochemistry and Biophysics*, *17*, 263-266.

Powell, J., White, I., Guy, D. & Brotherstone, S. (2008). Genetic parameters of production traits in Atlantic salmon (Salmo salar). *Aquaculture*, *274*, 225-231.

Pratoomyot, J., Bendiksen, E. Å., Bell, J. G. & Tocher, D. R. (2008). Comparison of effects of vegetable oils blended with southern hemisphere fish oil and decontaminated northern hemisphere fish oil on growth performance, composition and gene expression in Atlantic salmon (Salmo salar L.). *Aquaculture*, *280*, 170-178.

Quinton, C. D., McMillan, I. & Glebe, B. D. (2005). Development of an Atlantic salmon (Salmo salar) genetic improvement program: *Genetic parameters of harvest body weight and carcass quality traits estimated with animal models*, 211-217.

Raser, K. J., Posner, A. & Wang, K. K. W. (1995). Casein zymography - A method to study mu-calpain, m-calpain, and their inhibitory agents. *Archives of Biochemistry and Biophysics*, *319*, 211-216.

Rasmussen, R. S. (2001). Quality of farmed salmonids with emphasis on proximate composition, yield and sensory characteristics. *Aquaculture Research*, *32*, 767-786.

Regost, C., Jakobsen, J. V. & Rørå, A. M. B. (2004). Flesh quality of raw and smoked fillets of Atlantic salmon as influenced by dietary oil sources and frozen storage. *Food Research International*, *37*, 259-271.

Rosenlund, G., Obach, A., Sandberg, M. G., Standal, H. & Tveit, K. (2001). Effect of alternative lipid sources on long-term growth performance and quality of Atlantic salmon (Salmo salar L.). *Aquaculture Research*, *32*, 323-328.

Roth, B., Slinde, E. & Arildsen, J. (2006). Pre or post mortem muscle activity in Atlantic salmon (Salmo salar). The effect on rigor mortis and the physical properties of flesh. *Aquaculture*, *257*, 504-510.

Salem, M., Kenney, P. B., Killefer, J. & Nath, J. (2004). Isolation and characterization of calpains from rainbow trout muscle and their role in texture development. *Journal of Muscle Foods*, *15*, 245-255.

Salem, M., Kenney, P. B., Rexroad, C. E. & Yao, J. B. (2006). Molecular characterization of muscle atrophy and proteolysis associated with spawning in rainbow trout. *Comparative Biochemistry and Physiology D-Genomics & Proteomics*, *1*, 227-237.

Sanz, P. D., de Elvira, C., Martino, M., Zaritzky, N., Otero, L. & Carrasco, J. A. (1999). Freezing rate simulation as an aid to reducing crystallization damage in foods. *Meat Science*, *52*, 275-278.

Sargent, J. R., Tocher, D. R. & Bell, J. G. (2002). The lipids: Fatty acid peroxidation. In Fish Nutrition (3rd Edition). J.E. Halver, R.W. Hardy (Ed.). *Elsevier Science and Technology*, 500.

Sellers, A. & Murphy, G. (1981). Collagenolytic enzymes and their naturally occurring inhibitors. *International Review of Connective Tissue Research*, *9*, 151-190.

Sigholt, T., Erikson, U., Rustad, T., Johansen, S., Nordtvedt, T. S. & Seland, A. (1997). Handling stress and storage temperature affect meat quality of farmed-raised Atlantic salmon (Salmo salar). *Journal of Food Science*, *62*, 898-905.

Sikorski, Z. E. & Kolakowska, A. (1994). Changes in proteins in frozen stored fish, *Chapman & Hall*, London.

Sivertsvik, M., Jeksrud, W. K. & Rosnes, J. T. (2002). A review of modified atmosphere packaging of fish and fishery products - significance of microbial growth, activities and safety. *International Journal of Food Science and Technology*, *37*, 107-127.

Sivertsvik, M., Rosnes, J. T. & Kleiberg, G. H. (2003). Effect of modified atmosphere packaging and superchilled storage on the microbial and sensory quality of Atlantic salmon (Salmo salar) fillets. *Journal of Food Science*, *68*, 1467-1472.

Skjervold, P. O., Fjæra, S. O. & Østby, P. B. (1999). Rigor in Atlantic salmon as affected by crowding stress prior to chilling before slaughter. *Aquaculture*, *175*, 93-101.

Skjervold, P. O., Fjæra, S. O., Østby, P. B. & Einen, O. (2001a). Live-chilling and crowding stress before slaughter of Atlantic salmon (Salmo salar). *Aquaculture*, *192*, 265-280.

Skjervold, P. O., Fjæra, S. O., Østby, P. B., Isaksson, T., Einen, O. & Taylor, R. (2001b). Properties of salmon flesh from different locations on pre- and post-rigor fillets. *Aquaculture*, *201*, 91-106.

Skjervold, P. O., Rørå, A. M. B., Fjæra, S. O., Vegusdal, A., Vorre, A. & Einen, O. (2001c). Effects of pre-, in-, or post-rigor filleting of live chilled Atlantic salmon. *Aquaculture*, *194*, 315-326.

Sovik, S. L. & Rustad, T. (2006). Effect of season and fishing ground on the activity of cathepsin B and collagenase in by-products from cod species. *Lwt-Food Science and Technology*, *39*, 43-53.

Stien, L. H., Hirmas, E., Bjornevik, M., Karlsen, O., Nortvedt, R., Rørå, A. M. B., Sunde, J. & Kiessling, A. (2005). The effects of stress and storage temperature on the colour and texture of pre-rigor filleted farmed cod (Gadus morhua L.). *Aquaculture Research*, *36*, 1197-1206.

Suarez, M. D., Abad, M., Ruiz-Cara, T., Estrada, J. D. & Garcia-Gallego, M. (2005). Changes in muscle collagen content during post mortem storage of farmed sea bream (Sparus aurata): influence on textural properties. *Aquaculture International*, *13*, 315-325.

Suontama, J., Kiessling, A., Melle, W., Waagbo, R. & Olsen, R. E. (2007). Protein from Northern krill (Thysanoessa inermis), Antarctic krill (Euphausia superba) and the Arctic amphipod (Themisto libellula) can partially replace fish meal in diets to Atlantic salmon (Salmo salar) without affecting product quality. *Aquaculture Nutrition*, *13*, 50-58.

Suzuki, T. (1981). Fish and krill protein processing technology. *Appl. Sci. Publ. Ltd.*, London, 1-61.

Tardy, C., Codogno, P., Autefage, H., Levade, T. & Andrieu-Abadie, N. (2006). Lysosomes and lysosomal proteins in cancer cell death (new players of an old struggle). *Biochimica Et Biophysica Acta-Reviews on Cancer*, *1765*, 101-125.

Taylor, R. G., Geesink, G. H., Thompson, V. F., Koohmaraie, M. & Goll, D. E. (1995). Is Z-disk degradation responsible for postmortem tenderization? *Journal of Animal Science*, *73*, 1351-1367.

Taylor, R. G., Papa, I., Astier, C., Ventre, F., Benyamin, Y. & Ouali, A. (1997). Fish muscle cytoskeleton integrity is not dependent on intact thin filaments. *Journal of Muscle Research and Cell Motility*, *18*, 285-294.

Taylor, R. G. & Koohmaraie, M. (1998). Effects of postmortem storage on the ultrastructure of the endomysium and myofibrils in normal and Callipyge Longissimus. *Journal of Animal Science*, *76*, 2811-2817.

Taylor, R. G., Fjæra, S. O. & Skjervold, P. O. (2002). Salmon fillet texture is determined by myofiber-myofiber and myofiber-myocommata attachment. *Journal of Food Science*, *67*, 2067-2071.

Tejada, M. & Huidobro, A. (2002). Quality of farmed gilthead seabream (Sparus aurata) during ice storage related to the slaughter method and gutting. *European Food Research and Technology*, *215*, 1-7.

Thomas, P. M., Pankhurst, N. W. & Bremner, H. A. (1999). The effect of stress and exercise on post-mortem biochemistry of Atlantic salmon and rainbow trout. *Journal of Fish Biology*, *54*, 1177-1196.

Thomassen, M. S. & Røsjø, C. (1989). Different Fats in Feed for Salmon - Influence on Sensory Parameters, Growth-Rate and Fatty-Acids in Muscle and Heart. *Aquaculture*, *79*, 129-135.

Todorcevic, M., Kjær, M. A., Djakovic, N., Vegusdal, A., Torstensen, B. E. & Ruyter, B. (2008). Effects of dietary n-3 fatty acids on fat deposition, *mitochondrial function and oxidative stress in Atlantic salmon visceral adipose tissue*, XIII ISFNF. 1-5 June 2008, Florianopolis, Brasil.

Torstensen, B. E., Froyland, L., Ornsrud, R. & Lie, O. (2004). Tailoring of a cardioprotective muscle fatty acid composition of Atlantic salmon (Salmo salar) fed vegetable oils. *Food Chemistry*, *87*, 567-580.

Torstensen, B. E., Bell, J. G., Rosenlund, G., Henderson, R. J., Graff, I. E., Tocher, D. R., Lie, O. & Sargent, J. R. (2005). Tailoring of Atlantic salmon (Salmo salar L.) flesh lipid composition and sensory quality by replacing fish oil with a vegetable oil blend. *Journal of Agricultural and Food Chemistry*, *53*, 10166-10178.

Veiseth, E. & Koohmaraie, M. (2001). Effect of extraction buffer on estimating calpain and calpastatin activity in postmortem ovine muscle. *Meat Science*, *57*, 325-329.

Veiseth, E., Fjæra, S. O., Bjerkeng, B. & Skjervold, P. O. (2006). Accelerated recovery of Atlantic salmon (Salmo salar) from effects of crowding by swimming. Comparative

Biochemistry and Physiology Part B: *Biochemistry and Molecular Biology*, *144*, 351-358.

Vihko, V. & Salminen, A. (1983). Acid hydrolase activity in tissues of mice after physical stress. Comparative Biochemistry and Physiology Part B: *Biochemistry and Molecular Biology*, *76*, 341-344.

Wada, S., Hatai, K. & Kubota, S. S. (1991). Histopathological Study of Cultured Striped Jack with Yellow Fat Disease. *Fish Pathology*, *26*, 61-67.

Wang, M. Y. & Brown, W. D. (1983). Effects of elevated CO_2 atmosphere on storage of fresh-water crayfish (Pacifastacus leniusculus) *Journal of Food Science*, *48*, 158-162.

Wang, D., Tang, J., Correia, L. R. & Gill, T. A. (1998). Postmortem changes of cultivated Atlantic salmon and their effects on salt uptake. *Journal of Food Science*, *63*, 634-637.

Yamashita, M. & Konagaya, S. (1990). Participation of Cathepsin L into Extensive Softening of the Muscle of Chum Salmon Caught During Spawning Migration. *Nippon Suisan Gakkaishi*, *56*, 1271-1277.

Yamashita, M. & Konagaya, S. (1991). Hydrolytic Action of Salmon Cathepsin-B and Cathepsin-L to Muscle Structural Proteins in Respect of Muscle Softening. *Nippon Suisan Gakkaishi*, *57*, 1917-1922.

INTERNET SOURCES OF THE FIGURES:

http://www.earthlife.net/fish/images/anatomy/salmon-ts.jpg - Last access 26.04.2009

http://www.freerangefish.com/testsite/images/species/salmon_filet.jpg - Last access 27.0 4.2009

http://www.finecooking.com/assets/uploads/posts/5673/ING-salmon-filet-2_sql.jpg - Last access 09.09.2009

In: Salmon: Biology, Nutrition and Consumption
Editors: P. Lacopo et al., pp. 61-111
ISBN: 978-1-60741-287-8

Chapter 2

THE FEEDING HABITS OF PACIFIC SALMON, THEIR TROPHIC STATUS AND ROLE IN THE NORTH PACIFIC MARINE ECOSYSTEM

Svetlana V. Naydenko*
Pacific Scientific Research Fisheries Center (TINRO-Center),
4 Shevchenko Alley, Vladivostok 690600, Russia,

ABSTACT

Pacific salmon are among the dominant fish groups and the main consumers of forage resources in the upper layer of the subarctic Pacific. In recent years (1990s–2000s), the majority of Pacific salmon species (*Oncorhynchus* spp.) in North America and Asia experienced an increase in abundance and their role in the marine ecosystem has changed. In many publications the authors consider that the high Pacific salmon abundance increases interspecific food competition, and the density-dependent factors during periods of limited food availability are among the main reasons for the change in salmon production characteristics (length, weight, fecundity and increase in population average age).

The generalization of long-term data series (such as information on salmon and nekton feeding, the total quantitative estimates of nekton biomass and food supply for salmon—macroplankton and small nekton—along with consumption values of forage resources by salmon and other nekton species) integrated with Russian surveys carried out in the pelagic zone of the Far Eastern seas and adjacent waters of the northwestern Pacific Ocean made it possible to formulate a new perception of Pacific salmon status and their role in North Pacific pelagic communities.

Despite the fact that total Pacific salmon abundance increased significantly (concurrently with a decrease in large-size plankton biomass) in the Far Eastern marine ecosystems, there were quite a few cases in which it has impacted diet composition and size and interspecific food competition. Apparently, this is explained by the wide food plasticity of salmon (especially pink and chum salmon as the most abundant consumers), large standing stock of food resources, and some decrease in total nekton abundance.

* Email: naydenko@tinro.ru

In addition, Pacific salmon consume much less of the forage resources compared to many major species of nekton. For example, in the Pacific waters of the Kuril Islands, the consumption of forage resources by nekton (excluding Pacific salmon) during the summer period of 2004–2007 was 3.4–4.5 million tons, whereas Pacific salmon utilized 0.5–0.8 million tons. In the southern Okhotsk Sea in 1991 (in the period of high abundance of Pacific sardine *Sardinops melanosticta*), nekton species consumed 2.3 million tons of forage resources during summer, whereas Pacific salmon consumed as little as 0.5 million tons. In 2002–2003 (during the period of high abundance of juvenile walleye pollack *Theragra chalcogramma*), in the upper epipelagic layer of the western Bering Sea the overall amount of forage resources consumed by nekton species (excluding Pacific salmon) during autumn was 8.4–8.6 million tons, but Pacific salmon utilized as little as 1.0–1.3 million tons. In 2004–2006 (during the period of low walleye pollack abundance), the consumption of forage resources by nekton species dropped by 6–10 times. Probably such significant changes in abundance of the main consumers of forage resources (in this case walleye pollack and Pacific sardine) can change the food habits of Pacific salmon. The abundance of Pacific salmon has not increased significantly in comparison with most abundant nekton species and, therefore, Pacific salmon have not affected the normal functioning of pelagic communities.

Generally speaking, it is necessary to note that the Pacific salmon carrying capacity varies in different regions of the subarctic Pacific under the influence changing environmental factors.

INTRODUCTION

The last 25 years are characterized by heightened abundance of many Pacific salmon populations, including the Asian ones. Judging from anadromous migration activity (catch values, fish density on spawning grounds), the abundance became twice as high than in the previous period of low stocks in the 1950s–1970s (Shuntov et al. 1997; Temnykh et al. 2003, 2004; Shuntov and Temnykh 2004; Temnykh 2005; Shuntov 2005). In this connection, the problems of intra- and interspecies competition of salmon for the food resources are discussed in many scientific publications (Ishida et al. 1993, 2000; Bigler et al. 1996; Kaeriyama, 1996; Shiomoto at al. 1997; Watanabe 2000; Azumaya and Ishida 2000; Gritsenko and Klovatsh 2000; Bugaev 2001; Gritsenko et al. 2001; Mikodina et al. 2002; Ruggerone et al. 2003; Davis et al. 2004; Kaeriyama et al. 2004; Klovatsh 2004; Zavarina and Shevlyakov 2004; Yatsu and Kaeriyama 2005) with attempts to estimate the salmon abundance limitation by forage resources. It is concluded by some authors that the North Pacific is now overpopulated by salmon, that the trophic structure of its pelagic ecosystem has undergone fundamental reconstruction, and that the changes of production parameters noted recently for some salmon stocks (average length, weight, age and fecundity) are caused by density-dependent factors in conditions of limited food resources on marine feeding grounds. The opinion is claimed based on tough competition for the food not only among the salmon species but also with other consumers.

However, the extensive and detailed findings on the structure and dynamics of nekton and plankton communities, including their trophology, obtained during many complex large-scale surveys of the Far Eastern seas and northwest Pacific, summarized and published in Russian journals and monographs (Shuntov et al. 1993, 2007; Volkov et al. 1993; Volkov 1996; Shuntov 2001; Dulepova 2002; Nekton… 2003–2006; Temnykh 2004; Kuznetsova

2005; Chuchukalo 2006) allow the creation of a new conception of Pacific salmon status and their role in North Pacific pelagic communities.

Materials and Methods

The data on food habits of Pacific salmon and on plankton distribution and composition were collected on research vessels of the Pacific Fisheries Research Centre (TINRO) in the Okhotsk Sea, western Bering Sea and northwestern Japan Sea in 1986–2007, and on the R/V *TINRO, Sea Storm* and *Kaiyo maru* under the international research program BASIS (Bering-Aleutian Salmon International Survey) in the eastern Bering Sea in 2002–2007.

The surveys in the northwestern Japan, Okhotsk and western Bering seas were conducted using the standard midwater rope trawl with the headrope 80 m long and trawl net with the length 30 m, opening perimeter 396 m, quadrangular mesh in the body and wings and a small-meshed codend. The trawl hydrodynamic plate (6 m^2, 0.6 x 10 m) had floats on the headrope. The upper pelagic layer (0–50 m) was trawled.

Standard methods, developed earlier by Volkov and Chuchukalo (1986) and widely applied in TINRO-Centre, were used for sampling on feeding and processing the fish stomachs contents. The contents of all stomachs from the fish of the same species and similar size in each trawl catch had been combined in one sample that was processed with prey items identification and measurement of total prey weight and weights of all prey components. Daily food rations were calculated by the methods of Novikova (1949), Yurovitsky (1962), Kogan (1963), and Gorbatenko (1996). Total nekton consumption was determined as:

$$B=b*R*n,$$

where B – weight of the food consumed by nekton, tons; b – mean nekton biomass in certain period, tons; R – daily ration; n – duration of the period, days. Size-age, seasonal, year-to-year, and regional differences were taken into account in the calculations. The consumption of forage resources by nekton was calculated for the western Bering Sea in summer 2003 and autumn 2002–2006; for the southern Okhotsk Sea in summer 1991 and 2003 and autumn 2006; and for the Pacific waters of Kuril Islands in summer 1995, 2004, 2006 and 2007, because these years presented interest in terms of their comparison.

Plankton samples were collected with Juday net (nylon, mouth area 0.1 m^2, mesh size 0.168 mm) from the layers 0–50 and 0–200 m both in day and night just before trawling. The samples were divided into three size fractions: small-sized (animals < 1.2 mm in length), medium-sized (1.2–3.2 mm), and large-sized (> 3.2 mm). Species composition was determined for each fraction, with definition of weight (as displaced volume), size, and developmental stage for each species. The results on abundance and biomass were multiplied by catch ability coefficients (CC) of the Juday net (Table 1).

Table 1. Juday net catch ability coefficients for certain groups of plankton (from Volkov, 1996a)

Plankton size group	Catch ability coefficient (CC)
Small (< 1.2 mm)	1.5
Medium (1.2–3.2 mm)	2.0
Large (> 3.2 mm):	
euphausiids < 10 mm	2.0
euphausiids 10–20 mm	5.0
euphausiids > 20 mm	10.0
chaetognaths < 10 mm	2.0
chaetognaths 10–20 mm	5.0
chaetognaths > 20 mm	10.0
hyperiids < 5 mm	1.5
hyperiids 5–10 mm	5.0
copepods < 5 mm	2.0
copepods > 5 mm	3.0

All data were averaged by standard biostatistical regions proposed for the Russian EEZ by Shuntov et al. (1986; 1988a,b) and for the eastern Bering Sea by Volkov et al. (2004) (Figure 1).

Results and Discussion

Features of the Pacific Salmon Feeding in Different Parts of the Far Eastern Seas and Northwest Pacific

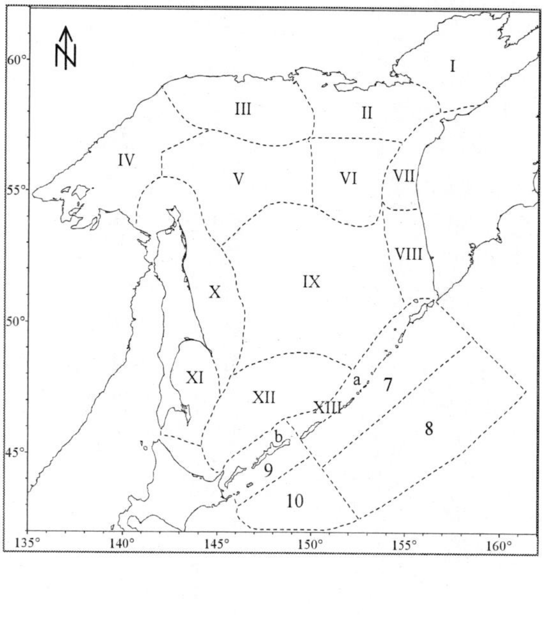

Figure Continued

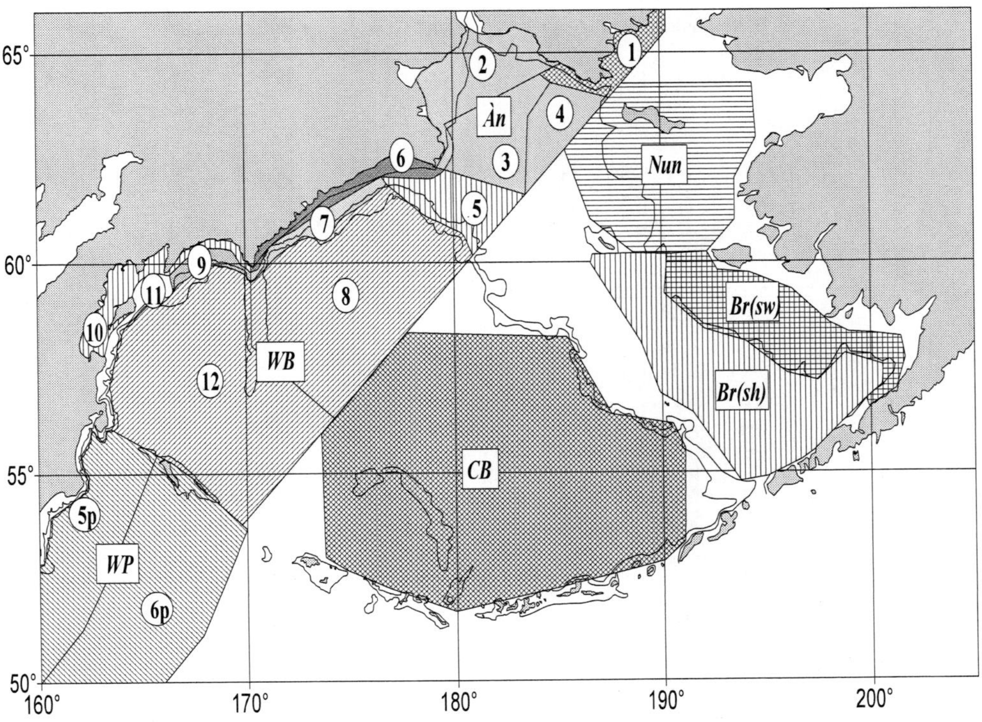

Figure 1. Biostatistical regions of data averaging: in the Okhotsk Sea (Fig. 1, upper left): I–VI, northern regions; VII–VIII, Western Kamchatka; IX, central Basin; X, eastern shelf of Sakhalin; XI, Terpenya Bay; XII, southern Basin; XIII a, northern and b, southern Kuril regions; 7–10, Pacific regions. In the northwestern Japan Sea (Fig. 1, upper right): 1, Tartar Strait; 2, northern Primorye; 3, southern Primorye; 4, deepwater area. In the Bering Sea (Fig. 1, lower panel): 1, Bering Strait; 2, northwestern Anadyr Bay; 3, southeastern Anadyr Bay; 4, eastern Anadyr Bay; 5, Navarin region; 6, Koryak shelf; 7, Koryak slope; 8, western Aleutian Basin; 9, Olutorskyi slope; 10, shelf of Karaginskyi and Olutorskyi bays; 11, Karaginskyi slope; 12, Commander Basin; 5p, Kamchatka Trench; 6p, oceanic waters off Kamchatka and Commander Islands; An, Anadyr Bay; WB, western Bering Sea basins; WP, oceanic waters near Commander Islands; CB, central deep Bering Sea; Nun, Nunivak region; Br(sw), shallow area of Bristol Bay (< 50 m); Br(sh), shelf zone of the Bristol Bay (> 50 m). (From Shuntov et al. 1986; Shuntov et al. 1988a, b; Volkov et al. 2004.)

Salmon species have specific types of feeding determined by their morpho-physiological features, but the rations composition depends on the structure of forage resorses, as well. By the type of feeding, the pink, chum, and sockeye salmon are typical euryphagous species which consume various plankton and nekton organisms, but the chinook and coho salmon are nekton consumers. The masu salmon, being euryphagous species by the type of feeding, is able to consume either plankton or nekton mainly in dependence on certain feeding grounds. Along with other factors, the specific type of feeding determines definite feeding habits of the salmon species. Following their feeding habits, certain species show preference for certain food, independently on the structure of plankton and nekton communities.

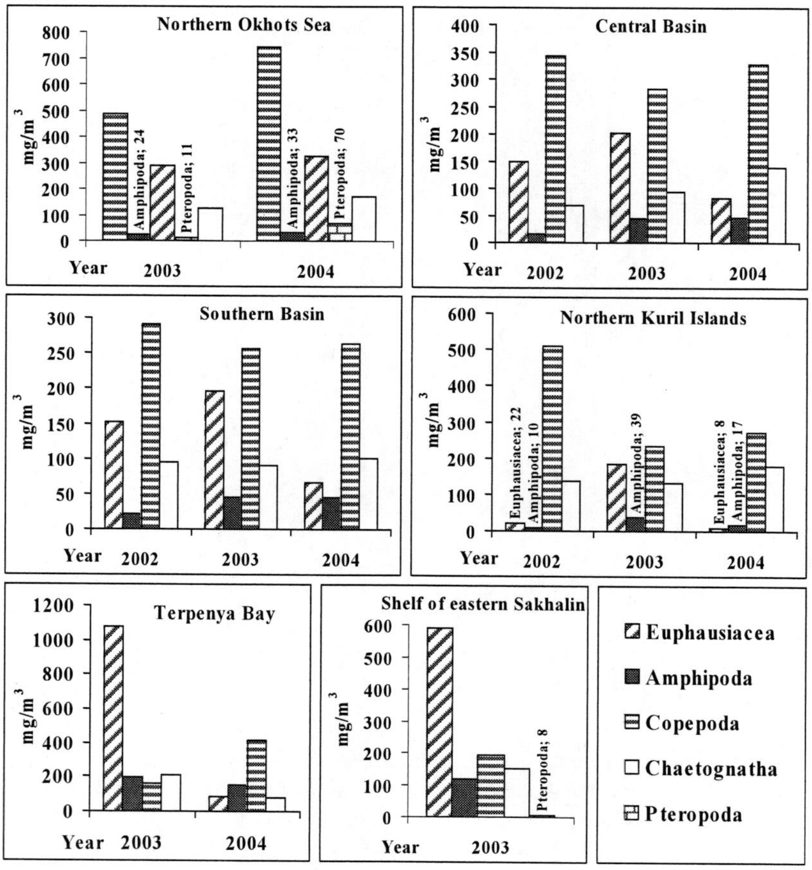

Figure 2. Biomass (mg/m^3) of some zooplankton groups in the epipelagic layer in the Okhotsk Sea in autumn, 2002–2004 (Naydenko et al. 2007).

Feeding of juvenile Pacific salmon

The juveniles of Pacific salmon prey on macroplankton in the period of migration to the open sea and extend their feeding on a variety of planktonic and nektonic organisms when they reach the open sea waters (Andrievskaya 1970, 1975; Gorbatenko and Chuchukalo 1989; Volkov 1996b; Gorbatenko 1996a; Karpenko 1998; Dulepova 1998; Lazhentsev and Bokhan 2001; Efimkin 2003; Efimkin et al. 2004; Volkov et al. 2006, 2007; Naydenko et. al 2007). Amphipods (hyperiids) and euphausiids were the predominant zooplankton prey in the diet of juvenile pink, chum, and sockeye during the whole period of our research, sometimes pteropod mollusks were important prey item, as well, but copepods and chaetognaths always had less importance. Amount of nektonic prey, as fish larvae and juveniles (walleye pollock, capelin, sand lances and other species) and juveniles of squids and crabs, varied significantly depending on season and region, with exception of juvenile coho and chinook salmon preyed predominantly on larvae and juveniles of different nektonic species. The diet of juvenile Pacific salmon depends on seasonal changes of prey species abundance and their vertical distribution in certain areas. The most considerable differences of their forage base are observed between the Japan, Okhotsk, and Bering seas.

Okhotsk Sea

Euphausiids (mainly Thysanoessa longipes, Th. inermis, Th. raschii, and to a lesser extent Euphausia pacifica), copepods (mainly Neocalanus plumchrus, Metridia pacifica, M. okhotensis, Oithona similis, Pseudocalanus minutus), and chaetognaths dominate in zooplankton in the most part of the Okhotsk Sea (Gorbatenko1996b; Shuntov 2001). For example, the area of high abundance of euphausiids (195–1077 mg/m^3) occupied the whole southern part of the Okhotsk Sea including the coastal waters of eastern Sakhalin and the Terpeniya (Patience) Bay in the autumns of 2002–2004; the high biomass of hyperiids (118–193 mg/m^3) was also found in this area (Figure 2).

The diets of juvenile pink and chum salmon are rather different in the Okhotsk Sea. Hyperiids (*Themisto pacifica, T. libellula* and *Primno macropa*) and euphausiids (*Th. longipes* and *E. pacifica*) are the main prey items for pink juveniles, which consume also calanoid copepods, pteropods, and chaetognaths in notable quantities (Figure 3), as is well-known from previous studies (Andrievskaya 1970; Gorbatenko and Chuchukalo 1989; Gorbatenko 1996a; Dulepova 1998; Lazhentsev and Bohan 2001; Naydenko and Kuznetsova 2002; Maksimenkov and Maksimenkova 2004; Nemchinova and Kovalenko 2004; Naydenko et al. 2007). Besides, the pink juveniles prey actively on young fish in shelf areas and sometimes on crab larvae (in particular in the years of its high abundance) in the northern part of the Sea and at the western Kamchatka (Gorbatenko and Chuchukalo 1989; Kuznetsova 2005). The main food objects for juvenile pink wintering in the southern Okhotsk Sea are hyperiids and euphausiids, and sometime squids larvae. For example, the shares of hyperiids and euphausiids in their diet were 53.1% and 31.9% in November–December of 1991, and hyperiids were 43.9–58.2%, euphausiids 17.3–26.3%, and squid larvae 11.3–20.0% in January, 1992 (Volkov 1996b; Gorbatenko 1996a).

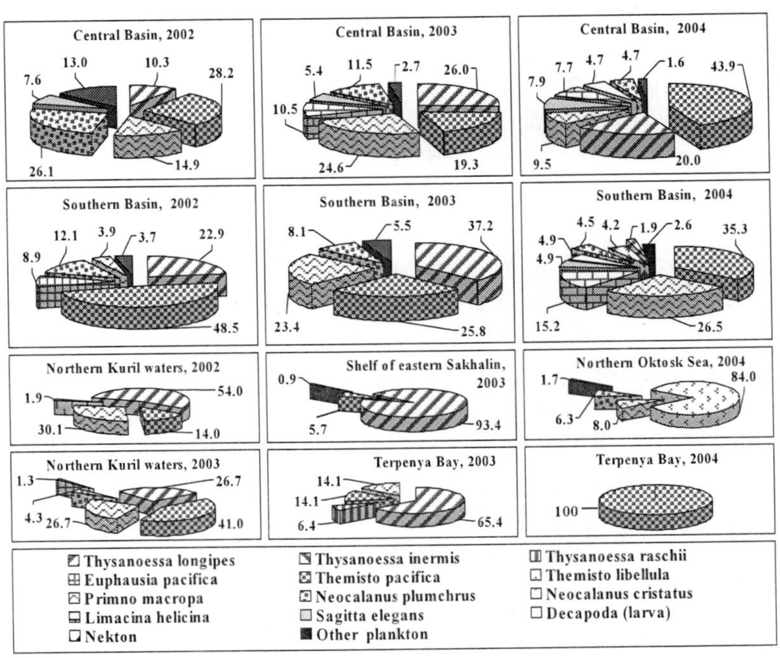

Figure 3. Diet (%) of juvenile pink salmon in the upper epipelagic layer in the Okhotsk Sea in autumn 2002–2004 (Naydenko et al. 2007).

Chum salmon juveniles consume mainly hyperiids (*T. pacifica* and *P. macropa*), euphausiids (*Th. longipes* and *E. pacifica*), calanoid copepods (mostly *N. plumchrus*), and chaetognaths in the southern Okhotsk Sea in summer and autumn, appendicularians are also presented in their diet in considerable quantity (see example in Figure 4). The same groups of plankton are typical for winter season, as well (Volkov 1996b; Gorbatenko 1996a). Young squids and fish (sand lance, pollock) are added to the chum juveniles diet in the northern part of the Sea and at western Kamchatka (Gorbatenko and Chuchukalo 1989; Kuznetsova 2005).

The diet of coho salmon juveniles includes hyperiids, copepods, pteropods, and young fish in the northern part of the Okhotsk Sea and larvae of crabs and shrimps in the waters at western Kamchatka (Maksimenkov and Maksimenkova 2004; Kuznetsova 2005; Chuchukalo 2006). Chinook salmon juveniles consume mainly squids and small fish, as well as euphausiids and crab larvae in the years of their high abundance (Glebov 2000; Maksimenkov and Maksimenkova 2004; Chuchukalo 2006). Feeding of masu salmon juveniles differs between the areas of the Okhotsk Sea, moreover, depends on seasons. For example, in August 1994 they preyed on hyperiids (*T. pacifica*) and copepods (*N. plumchrus* and *M. okhotensis*) in the central area, on young sand lance and euphausiids (*Th. longipes*) in the waters of Kuril Islands, on young pollock, herring, flounders, sand lance, capeling and crabs larvae on the shelf of Kamchatka, but pollock juveniles, crabs larvae, euphausiids, and copepods dominated in their diet at western Kamchatka in October, 2004 (Chuchukalo 2006) and young fish took 92% of their ration in the eastern part of the Sea in autumn 2001 (Maksimenkov and Maksimenkova 2004).

Western Bering Sea

Copepods (mainly *N. plumchrus, Calanus glacialis, M. pacifica, Eucalanus bungii, O. similis*) and chaetognaths are predominant groups of zooplankton in the upper pelagic layer of the western Bering Sea (Efimkin and Radchenko 1991; Shuntov et al. 1997; Efimkin et al. 2004; Volkov et al. 2004, 2005, 2007; Slabinskiy 2007). The copepods and chaetognaths are more abundant here than in the Okhotsk Sea. For example, copepod biomass was 163–1280 mg/m^3 at the coast and 75–499 mg/m^3 in the deep-water area; the biomasses of chaetognaths were 101–420 mg/m^3 and 124–289 mg/m^3, correspondingly, in fall seasons of 2002–2004. Pteropods and amphipods (mainly hyperiids) are more common in the Anadyr Bay and in the area at Cape Navarin (Figure 5).

In spite of copepods and chaetognaths domination in the zooplankton community, hyperiids, euphausiids and pteropods are the most important food items for juvenile pink and chum salmon in the western Bering Sea (Efimkin 2003; Efimkin et al. 2004a b; Maksimenkov and Smorodin 2004; Volkov et al. 2006; Chuchukalo and Kuznetsova 2006; Naydenko et al. 2007), at that hyperiids and pteropods often make the base of their ration in the Anadyr Bay and Cape Navarin region (Figure 6, 7). Besides these groups, fish larvae and juveniles are a considerable part of the diet, and copepods, chaetognaths, larvae of decapods, young squids, and sometimes other species are included into the ration in deep-water areas.

The dominant groups of the juvenile sockeye diet are usually copepods, early stages of euphausiids and hyperiids, crabs larvae, mysids, fish larvae and young fish in the shelf areas (Andrievskaya 1968; Burgner 1991; Chuchukalo et al. 1994; Volkov et al. 1997; Karpenko 1998; Chereshnev et al. 2002; Koval et al. 2007; Volkov et al. 2007), whereas hyperiids, euphausiids, and copepods in the deep-water areas (data for fall seasons of 2002–2007), with exception of the Commander Basin where the diet includes pteropods and chaetognaths, as well. The last area is curious by both heightened pteropods abundance in plankton and their higher share in juvenile salmon ration, in compare with the western Aleutian Basin (Figure 8). So, feeding of the pink, chum, and sockeye juveniles has spatial change in the western Bering Sea depended on distance from the shore; similar changes are found in the northwest Pacific off the east coast of Kamchatka (Koval et al. 2007).

The diet of chinook juveniles includes the larvae and juveniles of squids (mostly *Gonatus kamtschaticus*, *Gonatopsis borealis*) and fishes (capelin, sand lance, pollock) in the whole western Bering Sea (Figure 9). Coho juveniles have a similar diet, but fish dominate in their ration in the Commander Basin, and squids – in the western Aleutian Basin (Efimkin 2003; Efimkin et al. 2004a,b; Maksimenkov and Smorodin 2004; Volkov et al. 2006; Chuchukalo Kuznetsova 2006; Volkov et al. 2007; Naydenko 2007; Koval et al. 2007) (Figure 10).

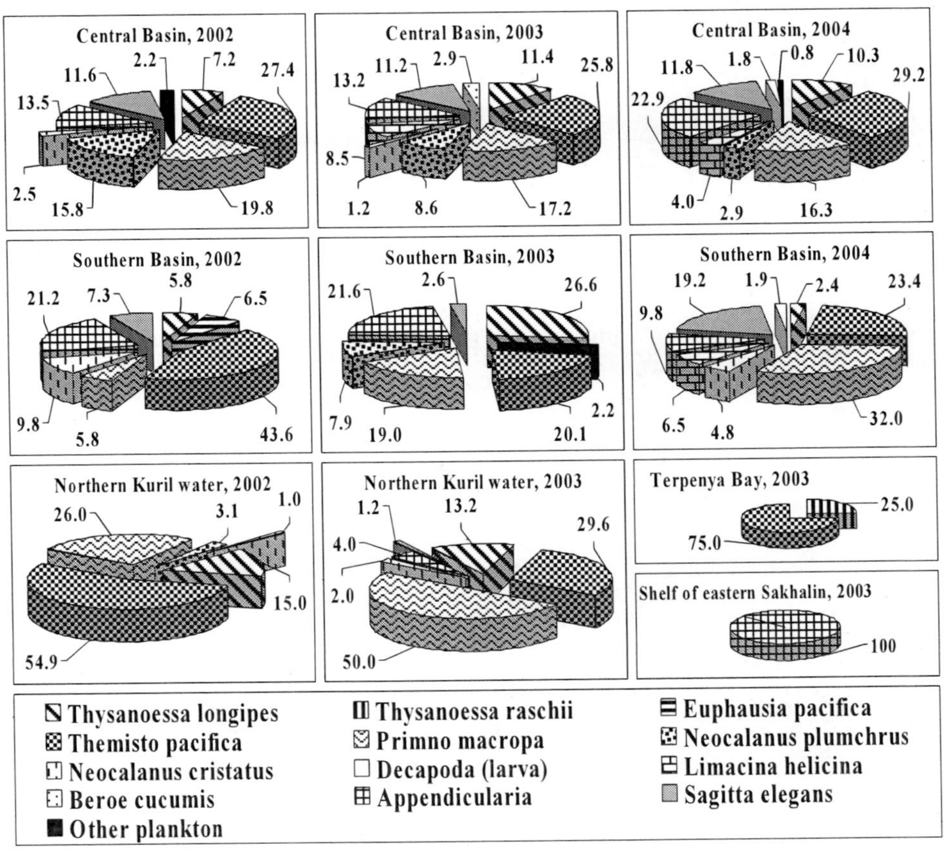

Figure 4. Diet (%) of juvenile chum salmon in the upper epipelagic layer in the Okhotsk Sea in autumn 2002–2004.

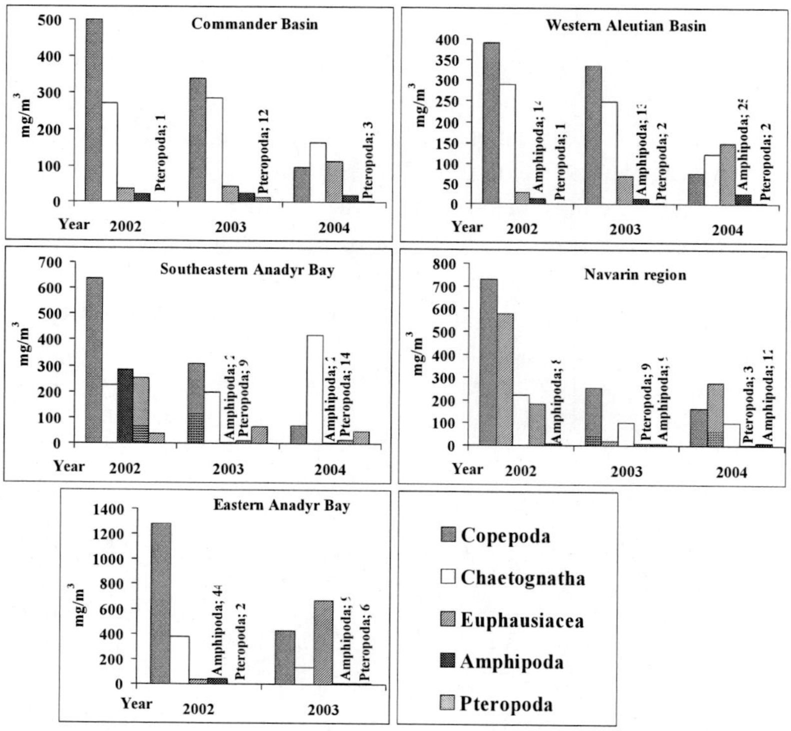

Figure 5. Biomass (mg/m^3) of some groups of zooplankton in the epipelagic layer in the western Bering Sea in autumn 2002–2004 (Naydenko et al. 2007).

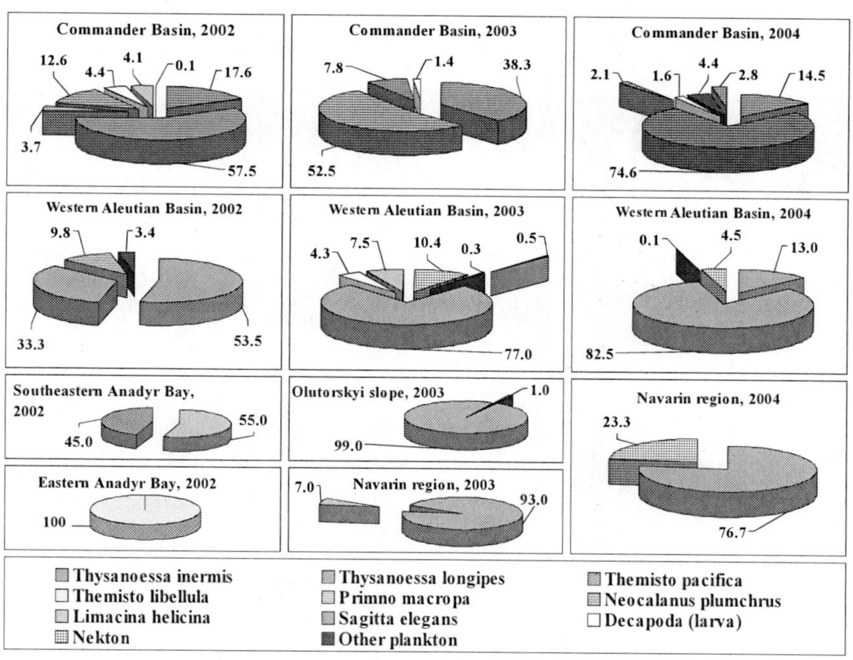

Figure 6. Diet (%) of juvenile pink salmon in the upper epipelagic layer in the western Bering Sea in autumn 2002–2004 (Naydenko et al. 2007).

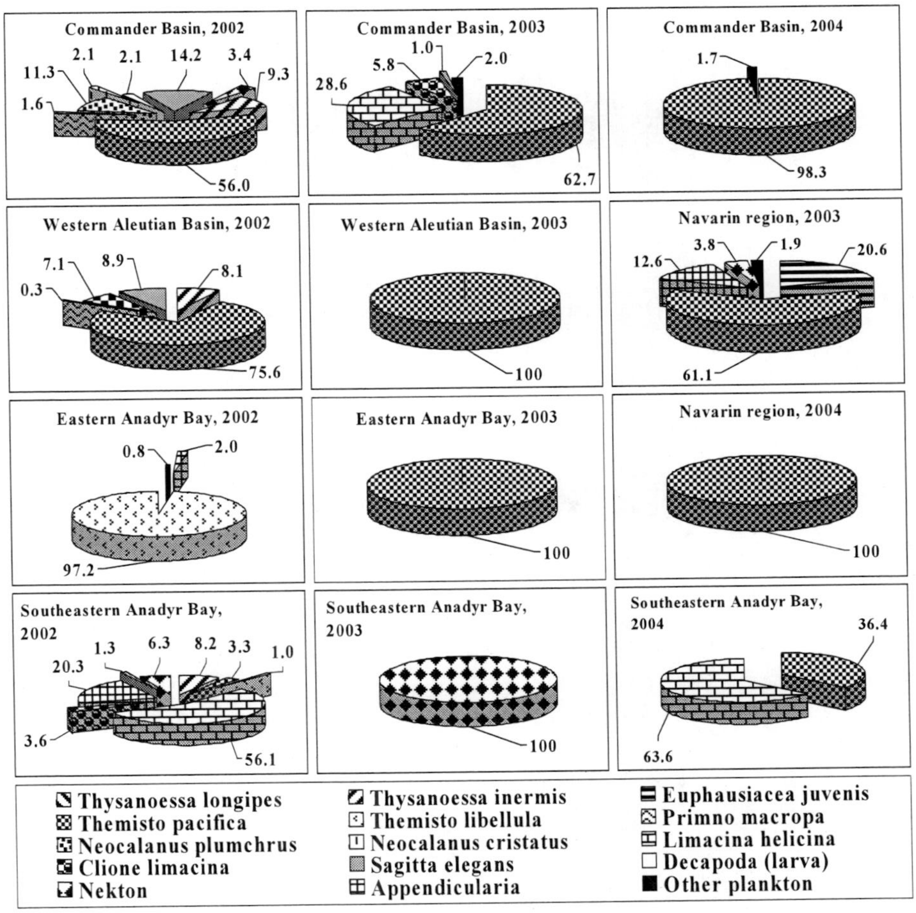

Figure 7. Diet (%) of juvenile chum salmon in the upper epipelagic layer in the western Bering Sea in autumn 2002–2004.

Eastern Bering Sea

Ichthyoplankton and meroplankton species dominate in plankton of the eastern Bering Sea in autumn of 2000s (Efimkin et al. 2004; Volkov et al. 2004; Volkov et al. 2005; Volkov et al. 2007). Juvenile pink and chum salmon prey mainly upon larvae and juveniles of walleye pollock and capelin and upon larvae of crabs and bottom fishes (data of the surveys in the Bristol Bay shelf and in the area at Ninivak Island obtained in September–October of 2003–2007) (Figure 11, 12, 13). The diet of juvenile chum is wider and includes also gelatinous zooplankton (medusas and appendicularians *Oikopleura vanhoeffeni*). The diet of juvenile sockeye in both areas is based on fish (mostly larvae and juveniles of pollock and capelin), other groups as euphausiids, hyperiids, and mysids are presented in a lesser degree (Figure 8). The juveniles of chinook and coho consume fish mainly, but crabs larvae are important for them in the Nunivak region, as well (Efimkin et al. 2004; Volkov et al. 2006; Kuznetsova 2006; Kuznetsova et al. 2007) (Figure 9, 10).

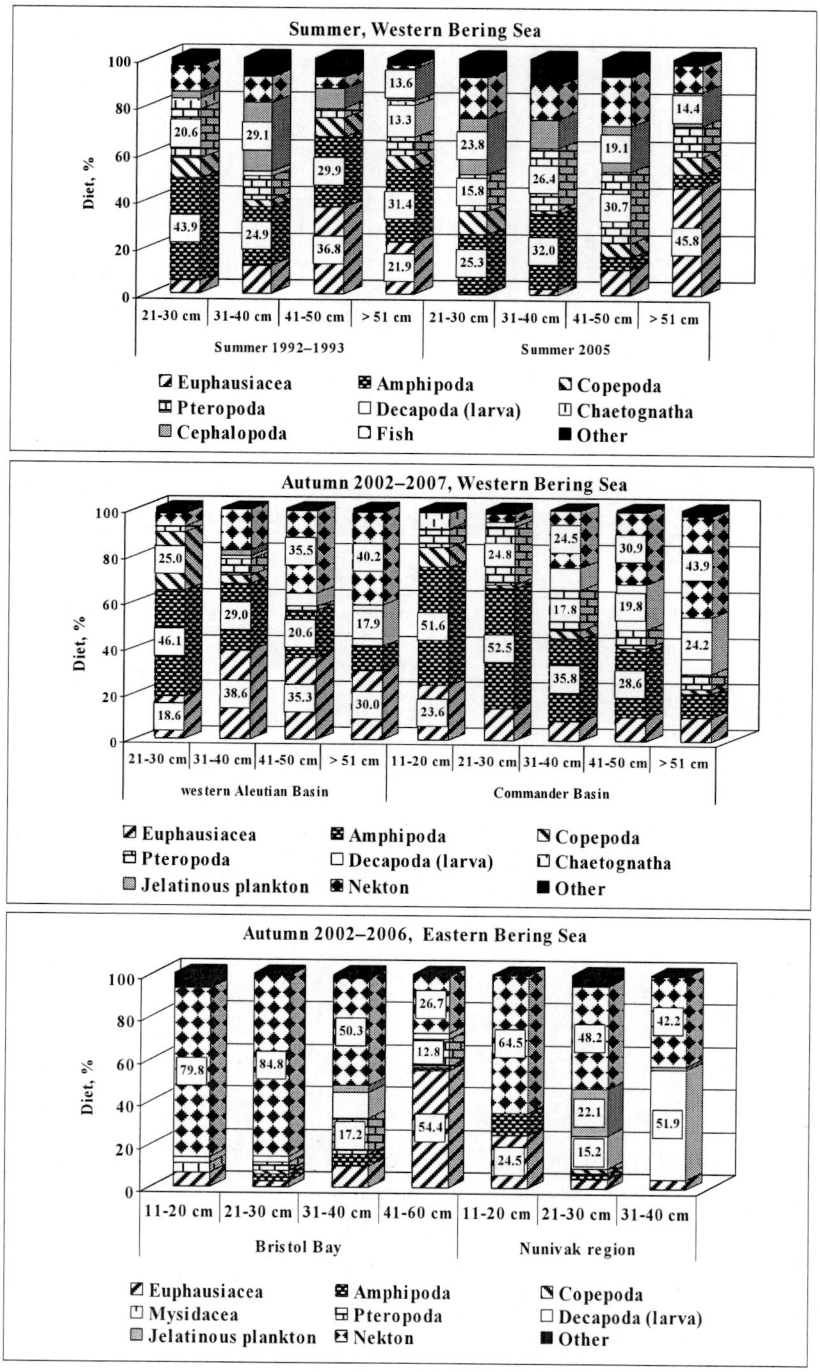

Figure 8. Diet (%) of sockeye in the upper epipelagic layer in summer and autumn in the western Bering Sea and in autumn in the eastern Bering Sea (from Chuchukalo et al. 1994, Volkov et al. 1997; Efimkin et al. 2004; Chuchukalo 2004; Volkov et al. 2007 with modifications).

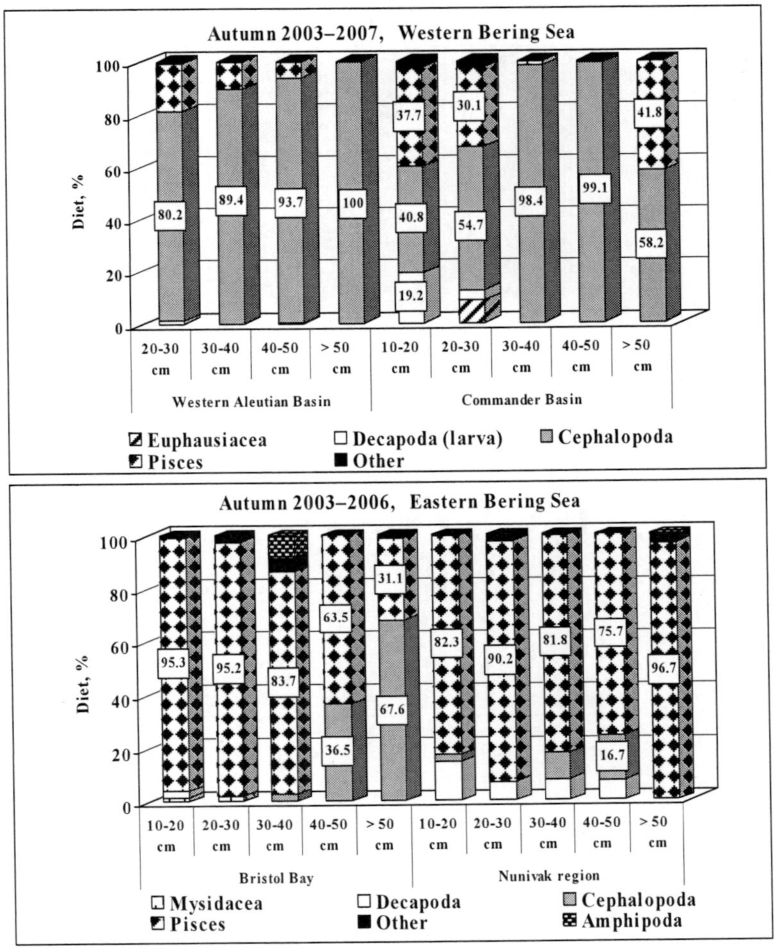

Figure 9. Diet (%) of chinook in the upper epipelagic layer in the western Bering Sea in autumn 2003–2007 and in the eastern Bering Sea in autumn 2003–2006 (from Efimkin et al. 2004; Volkov et al. 2007; with modifications).

Northwest Pacific

Feeding of salmon in the Northwest Pacific was investigated in detail in winter and spring of 1989 and 2006. In February 1989, the base of pink salmon juveniles diet was formed by Myctophidae fishes at night, when they rised to the sea surface, and hyperiids, chaetognaths, and salps in the daytime. In March 1989, copepods (mostly *Neocalanus cristatus*) and appendicularians dominated, and hyperiids, chaetognaths, and euphausiids were presented in the diet in a lesser degree. In May 1989, chaetognaths and copepods (*N. cristatus*) dominated (Tutubalin and Chuchukalo 1992). The ration of chum salmon juveniles was based on hyperiids, copepods, chaetognaths, and salps (in decreasing order) during the whole period of survey in 1989 (Chuchukalo 2006). In winter of 2006, the diets of all salmon were very diverse and distinguished by high spatial variability, but the shares of hyperiids, euphausiids, chaetognaths, and copepods were usually the most considerable. Besides these groups, squids were important for the ration of chum, chinook, and coho juveniles, and fish for the ration of sockeye (Volkov 2006; Volkov et al. 2007).

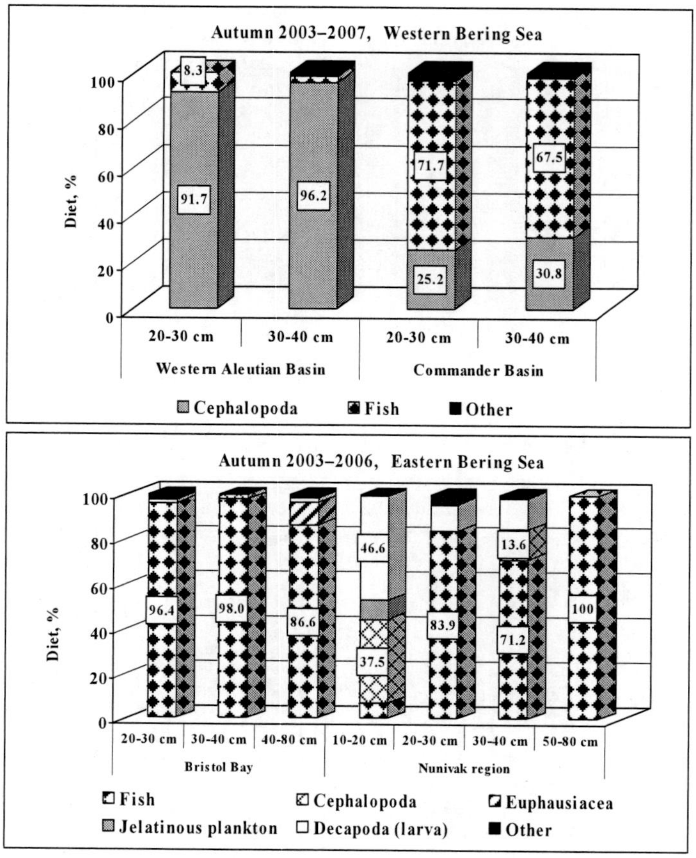

Figure 10. Diet (%) of coho in the upper epipelagic layer in the western Bering Sea in autumn 2003–2007 and in the eastern Bering Sea in autumn 2003–2006 (from Efimkin et al. 2004; Volkov et al. 2007; with modifications).

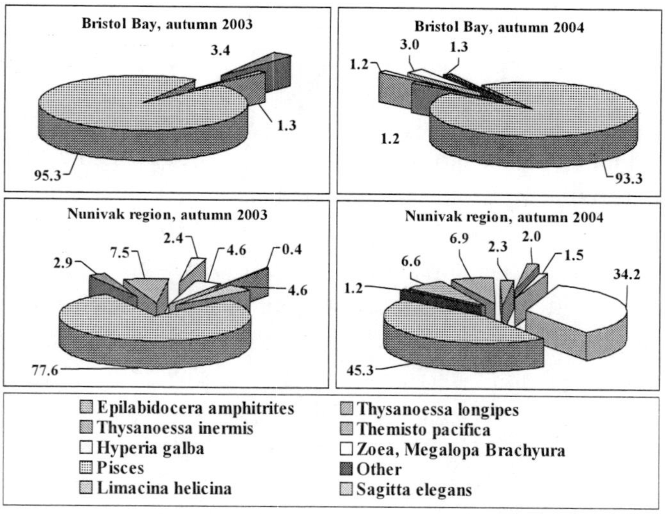

Figure 11. Diet (%) of juvenile pink salmon in the upper epipelagic layer in the eastern Bering Sea in autumn 2003–2004 (from Efimkin et al. 2004; Volkov et al. 2007).

Northwestern Japan Sea

Copepods (mainly *C. glacialis, M. pacifica, M. okhotensis, N. cristatus, N. plumchrus*), chaetognaths, and euphausiids (*Th. longipes, Th. inermis, E. pacifica*) comprised the bulk of the plankton community in the northwestern Japan Sea (Dolganova 2001; Shuntov 2001). Hyperiids (*T. pacifica*) are more abundant here than in the Okhotsk and Bering seas, but pteropods (*Limacina helicina*) are less abundant. However, the share of *T. pacifica* in the juvenile pink salmon diet is lower in the Japan Sea because the juveniles prey here additionally upon another hyperiid species (*P. macropa*) as well as upon euphausiids, highly abundant in the upper pelagic layer of the northwestern Japan Sea (Naydenko et al. 2007) (Figure 14).

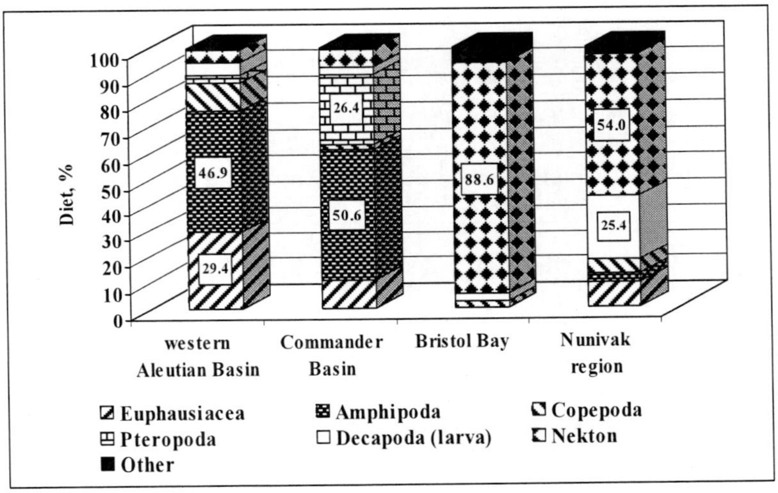

Figure 12. Diet (%) of juvenile pink salmon in the upper epipelagic layer in the western and eastern Bering Sea in autumn 2003–2007 (from Efimkin et al. 2004; Volkov et al. 2007 with modifications).

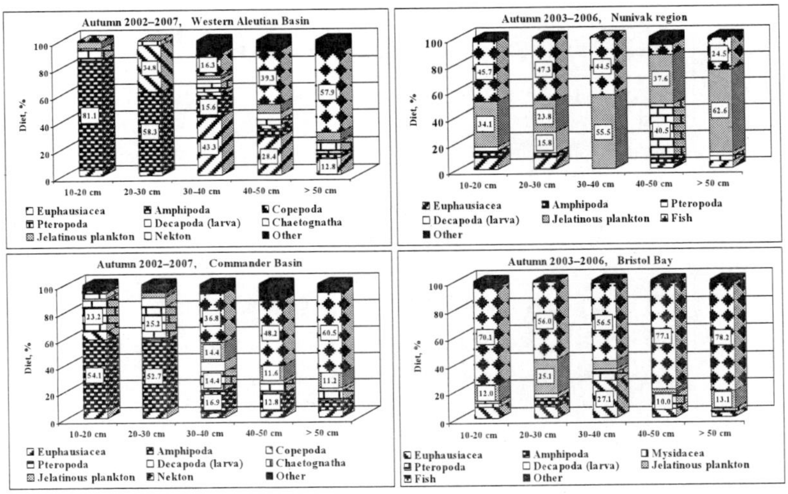

Figure 13. Diet (%) of chum salmon in the upper epipelagic layer in the Bering Sea in autumn 2002–2007 (from Efimkin et al. 2004; Volkov et al. 2007 with modifications).

As the summary of studies in different seas, there is concluded that the diets of pink, chum, and sockeye juveniles are formed mainly by euphausiids, hyperiids, copepods, and pteropods, though nekton groups as young fish and larvae of crabs and shrimps can be important in coastal areas. The diets of coho, chinook, and masu juveniles include nekton predominantly. The share of different groups in the rations could be changed in dependence on the state of forage resorses in certain regions.

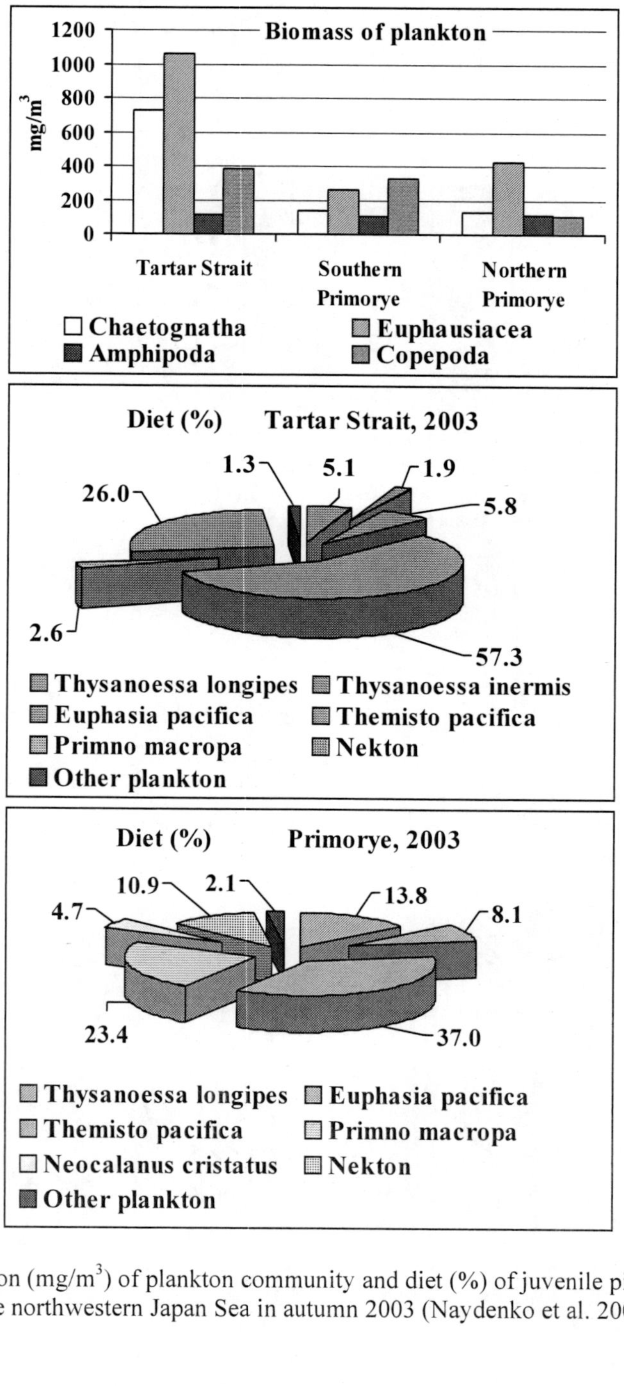

Figure 14. Composition (mg/m^3) of plankton community and diet (%) of juvenile pink salmon in the epipelagic layer in the northwestern Japan Sea in autumn 2003 (Naydenko et al. 2007).

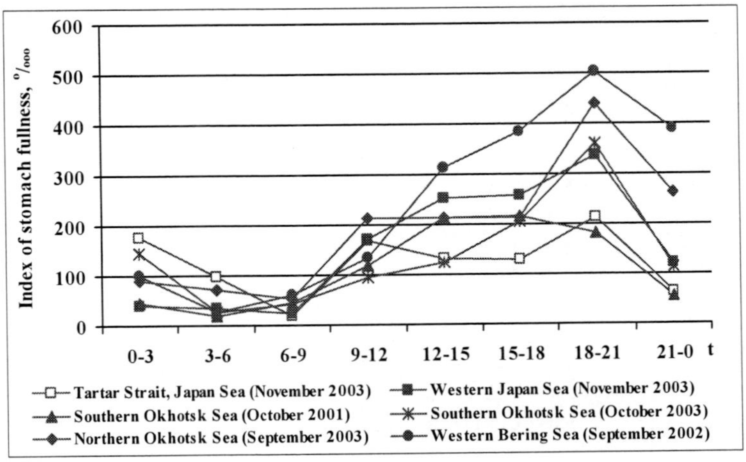

Figure 15. Daily rhythm of juvenile pink salmon feeding in the Far Eastern seas in autumn (Naydenko et al. 2007).

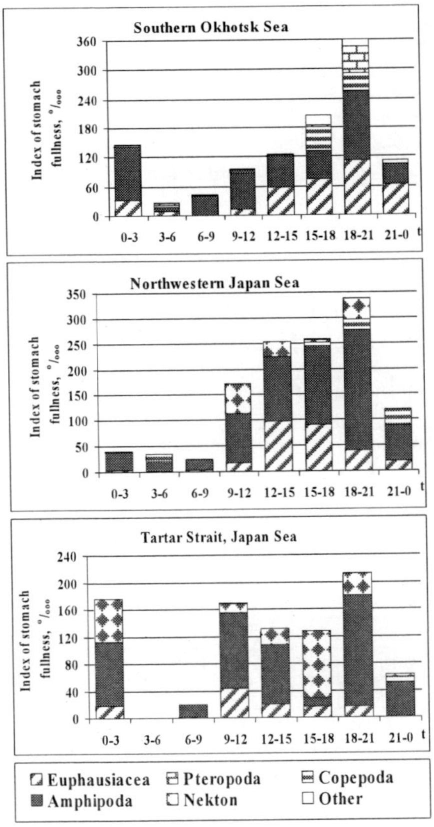

Figure 16. Consumption of some zooplankton groups by juvenile pink salmon during day and night in the southern Okhotsk and northwestern Japan Seas in autumn 2003.

Table 2. Daily diets of juvenile Pacific salmon

Region	Data	Daily diets, %	References
1	2	3	4
Juvenile pink			
Central Okhotsk Sea	September–October 1985	5.5–7.2	Gorbatenko and Chuchukalo 1989
Southern Okhotsk Sea	October 1998 October 1999	4.7 4.0–4.4	Lazhentsev and Bohan 2001
Northern Okhotsk Sea	September–October 1994 September–October 1998 September–October 2000	8.5–14.2 8.5 7.5	Kuznetsova 2005
	October–November 2007	8.6	unpublished data of A.Ya. Efimkin and N.S. Kosenok, TINRO-Centre
Southern Okhotsk Sea	October–November 2007	5.0–8.2	
	November–December 1991	1.4–1.9	Lazhentsev and Bohan 2001
	November–December 1994 November–December 1995	2.6 4.7	Dulepova 1998
	January 1992	1.7–4.5	Gorbatenko 1996
Tartar Strait (Japan Sea) Primorye (Japan Sea)	November 2003	6.9 9.3	unpublished data of A.E. Lazhentsev, TINRO-Centre
Northwest Pacific	February 1989 March 1989 April–May 1989	5.0 10.0 7.0	Tutubalin and Chuchukalo 1992; Chuchukalo 2006
Juvenile chum			
Southern Okhotsk Sea	June–July 2003	7.6–9.9	Nemchinova and Kovalenko 2004
Central Okhotsk Sea	September–October 1985	4.2–5.0	Gorbatenko and Chuchukalo 1989
Northern Okhotsk Sea	September–October 1994 September–October 1998	7.3 6.8	Kuznetsova 2005
Okhotsk waters of western Kamchatka	September–October 1986	6.1	Gorbatenko and Chuchukalo 1989
Western Bering Sea	September–October 1986–1990 September–October 2002 September–October 2003 September–October 2004	6.3–8.4 7.5 7.0 7.9	Efimkin et al. 2004; Shuntov and Temnykh 2004; Chuchukalo 2006
Southern Okhotsk Sea	November–December 1991	2.2–2.7	Lazhentsev and Bohan 2001
	January 1992	2.9	Gorbatenko 1996
	November–January 1994–1995	3.5	Naydenko and Kuznetsova 2002
	November–January 1995–1996	2.6	
	October–November 2007	5.0–6.4	unpublished data of A.Ya. Efimkin and N.S. Kosenok, TINRO-Centre
Northern Okhotsk Sea	October–November 2007	7.7	
Region	**Data**	**Daily diets, %**	**References**
1	2	3	4
Northwest Pacific	February–March 1989 April–May 1989	2.5–3.0 5.0 a	Tutubalin and Chuchukalo 1992; Chuchukalo 2006
Juvenile sockeye			
Western Bering Sea	June–July 1992 June–July 1993	4.1 0.8–2.5	Volkov et al. 1997; Chuchukalo 2006

Pacific waters of eastern Kamchatka	June–August 1993	1.9	
Western Bering Sea	September–October 1992 September–October 2002 September–October 2003 September–October 2004	4.1 4.7 6.0 6.0	Efimkin et al. 2004; Shuntov and Temnykh 2004
Juvenile chinook			
Western Bering Sea	October 1987 November–December 1991	3.8 3.9	Glebov 2000
Okhotsk waters of western Kamchatka	August 1996 September 1994 November 1991	4.4 4.8 4.2	Glebov 2000
Pacific waters of Kuril Islands	November 1987 November 1991	4.4 4.3	Glebov 2000
Juvenile coho			
Central Okhotsk Sea	June–August 1991	3.0	Chuchukalo 2006
Pacific waters of Kuril Islands	June–August 1991 June–August 1993	3.4 3.6	
Pacific waters of eastern Kamchatka	June–August 1993	2.1	
Okhotsk waters of western Kamchatka	June–August 1992	3.3	
Okhotsk Sea	September–October 1998	4.5–7.2	
Northern Okhotsk Sea	September–October 1998	5.1	Kuznetsova 2005
Western Bering Sea	September–October 2002 September–October 2004	4.4 4.2	Efimkin 2003
Juvenile masu			
Okhotsk waters of western Kamchatka	August–September	2.4	Chuchukalo 2006

Diurnal rhythm of feeding is distinguished by the highest activity in day and evening times and the lowest one at night for the juveniles of all salmon species. These features are determined, in certain degree, by diurnal changes of food ability in the layer of the salmon dwelling (Tutubalin and Chuchukalo 1992; Volkov 1996b; Gorbatenko 1996a; Lazhentsev and Bohan 2001; Nemchinova and Kovalenko 2004; Maksimenkov and Maksimenkova 2004; Koval et al. 2007; Naydenko et al. 2007). For example, the share of indigested crustaceans (euphausiids, copepods, hyperiids) in the fish stomachs collected in deep-water areas of the Okhotsk and western Bering seas increases considerably in 18–22 p.m., when these plankton animals rise to the upper layer (Figure 15). The highest activity of the pink salmon juveniles feeding by fish larvae and decapods in the Tartar Strait (Japan Sea) is observed in the daytime (Naydenko et al. 2007) (Figure 16), similar rhythm of feeding is noted for other species of salmon juveniles, as well (Tutubalin and Chuchukalo 1992; Volkov 1996b; Gorbatenko 1996a; Lazhentsev and Bohan 2001; Nemchinova and Kovalenko 2004; Maksimenkov and Maksimenkova 2004; Koval et al. 2007; Naydenko et al. 2007).

Seasonal rhythm of the salmon juveniles feeding consists in periods of high feeding activity in summer-autumn and spring, and usually lowered activity in winter (Table 2). For instance, pink juveniles have the daily ration 8.5–14.2% of their body weight in summer and autumn that decreases to 1.4–5.0% in December–January and rises again to 7.0–10.0% in spring (Table 2). However, the daily rations can vary significantly within the same regions and seasons presumably because of variations of forage resources state, prey availability and its food value.

Feeding of Pacific Salmon on Feeding Grounds and in Anadromous Migration

Pink salmon

The diet of pink salmon includes plankton and small-sized nekton (Heard 1991; Andrievskaya 1975; Prahova et al. 1980; Volkov et al. 1997; Kuznetsova 2005; Chuchukalo 2006; and others). Usually the share of planktonic organisms is higher in the Okhotsk Sea and in the Pacific waters of Kuril Islands than in the Bering Sea, in particular in the southern part of the Okhotsk Sea, where the share of euphausiids was 40–49% in summer periods of 1991–2003. Exactly in this area the euphausiids abundance in plankton is the highest. Hyperiids have usually the heightened abundance in the shelf of western Kamchatka, and in the Central and in the northern part of the Sea, conformably their share in the pink salmon ration is higher in these areas and reaches 34–54% (Figure 17). Small nekton is presented in the pink salmon ration by young squids, larvae and juveniles of pollock, herring, and atka mackerel, adults of capelin and sand lance in coastal areas of the Okhotsk Sea and by mesopelagic fish species, as myctophids and northern smoothtongue at the continental slope (Prahova et al. 1980; Gorbatenko and Chuchukalo 1989; Chuchukalo et al. 1992a, b, c; Gorbatenko 1996a; Volkov et al. 1997; Naydenko and Kuznetsova 2002; Kuznetsova 2005). Euphausiids are the main food for pink salmon in coastal waters of southern Kuril Islands, but its diet in the Pacific waters of northern Kuril Islands and oceanic waters of southern Kurils are based on three planktonic groups (euphausiids, pteropods, and copepods), and nekton in a lesser degree (data for summer 2004–2007) (Figure 18). Similar data was received for summer 1991–1996. Nekton food was presented in the pink salmon diet small saury and japanese anchovy in these regions (Chuchukalo et al. 1992b; Naydenko and Kuznetsova 2002). Crabs and shrimps larvae, small fish and squids are a considerable portion of the ration of pink salmon in deep-water areas of the western Bering Sea, in particular in the western Aleutian Basin, but 28–52% of the ration in the Commander Basin are composed by plankton: hyperiids, euphausiids, copepods (data for summer 1991–1995, 2003). The crabs and shrimps larvae and juveniles of flounders and sculpins are the base of the pink salmon diet in the shelf areas at Navarin Cape and at the eastern coast of Kamchatka (Chuchukalo et al. 1992c; Volkov 1994; Volkov et al. 1997). Planktonic groups become dominant in the diet of pink salmon in winter and spring (Heard 1991; Tutubalin and Chuchukalo 1992; Volkov et al. 2007). The joint share of chaetognaths, salps, hyperiids, and copepods reaches 61–85% in winter (by the data of 1989 and 2006), and increased to 92–100% in spring. In winter the rest of the ration is presented by Myctophidae species. In general, although the pink salmon diet changes considerably in dependence of forage resorses of certain area, this species prefers usually hyperiids (*T. pacifica, P. macropa*), euphausiids (*Th. longipes, Th. inermis, E. pacifica*), pteropods, and larvae of crabs and shrimps.

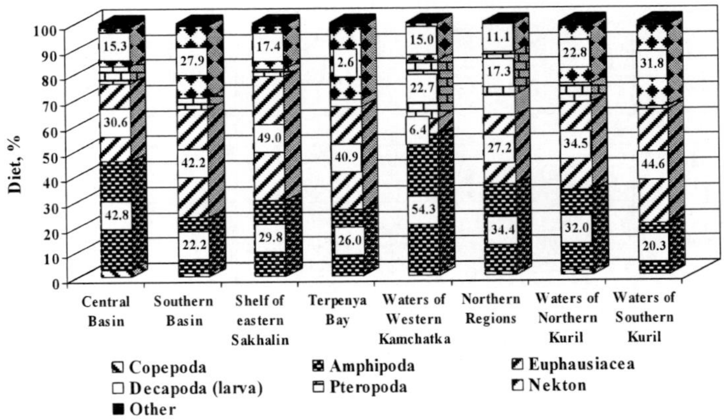

Figure 17. Diet (%) of pink salmon in the different regions of Okhotsk Sea in summer 1991–2003 (from Volkov 1994; Volkov et al. 1997; Kuznetsova 2005; Chuchukalo 2006; with modifications and additions).

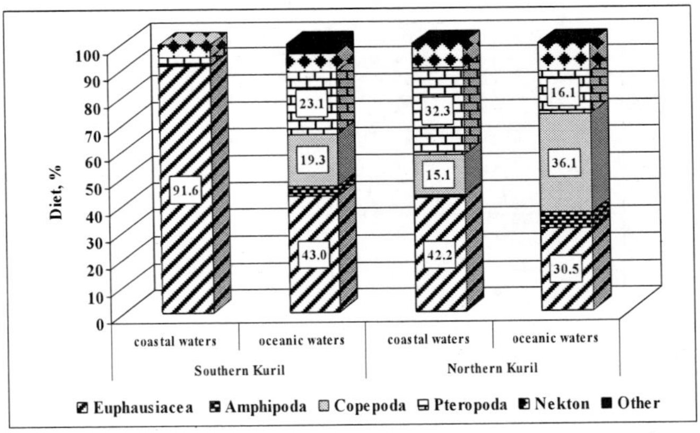

Figure 18. Diet (%) of pink salmon in the Pacific waters of the Kuril Islands in summer 2004–2007.

Feeding activity of pink salmon in the feeding period is rather high. Its daily ration is estimated as 5–10% from the body weight for the winter-spring feeding in the ocean waters (Tutubalin and Chuchukalo 1992). The species continues active feeding in its anadromous migration when daily ration sometimes reaches the values 4.5–8.8% (Chuchukalo et al. 1992a,b,c; Volkov 1994; Naydenko and Kuznetsova 2002; Kuznetsova 2005; Chuchukalo 2006). However, the feeding activity decreases gradually as migration goes and fish approach the spawning grounds; the daily ration decreases to 2.5–0.5% when the gonad-somatic index is 12–15%, and the ration value falls to the minimal value of 0.5% when the gonad-somatic index is 15–16% (Kuznetsova 2005; Chuchukalo 2006).

Chum salmon

The diet composition and feeding activity of chum salmon is rather variable spatially and temporary and in dependence of size and age of the fish. (Andrievskaya 1975; Salo 1991; Starovoytov 2003; Chuchukalo 2006; Volkov et al. 2007).

Table 3. Daily diets of chum salmon

Region	Data	Length, cm	Daily diets, %	References
1	2	3	4	5
Western Bering Sea	June–July 1991	50–60 > 60	4.6 2.2	Starovoytov 2003
	June–July 1993	40–50 50–60 > 60	2.5 2.4 1.6	Volkov et al. 1997; Starovoytov 2003
	June–July 1995	50–60 > 60	2.0 2.0	Starovoytov 2003
	September–November 1986	30–40 > 50	3.2 2.0	Shuntov and Temnykh 2004
	September–November 1987	30–40	3.8	
	September–October 2002	30–40 40–50 50–60	4.6 3.8 4.5	Efimkin et al. 2004; Zavolokin et al. 2007
	September–October 2003	30–40 40–50 50–60 > 60	3.9 3.9 3.8 1.0	
	September–October 2004	30–40 40–50 50–60	2.1 2.4 3.2	Shuntov and Temnykh 2004; Zavolokin et al. 2007
Navarin region (Western Bering Sea)	August 2004	30–40 40–50 50–60	3.8 2.6 1.8–2.6	Kosenok and Naydenko 2008
Western Aleutian Basin (Western Bering Sea)	August 2004	30–40 40–50	2.6 2.8	
	September 2004	30–40 40–50	2.6 2.8	
Pacific waters of eastern Kamchatka	June–July 1991	50–60 > 60	3.4–4.4 2.5	Starovoytov 2003
	June–July 1993	30–40 40–50 50–60 > 60	1.6 1.6 2.0 1.3	Volkov et al. 1997
	September-October 2004	50–60 > 60	1.3 1.3	unpublished data of N.S. Kosenok, TINRO-Centre
	June–July 1994	50–60	4.7	Starovoytov 2003
	June–July 1995	50–60 > 60	1.5 1.4	
Pacific waters of Kuril Islands	June–July 1991	50–60	5.6	
	June–July 1993	30–40 40–50 50–60 > 60	3.1 2.2 1.2–2.1 1.3–1.5	Volkov et al. 1997
Region	**Data**	**Length, cm**	**Daily diets, %**	**References**
1	2	3	4	5
Open Pacific waters of Northern Kuril	June 2007	40–50 50–60	3.5 3.8	unpublished data of A.E. Lazhentsev, TINRO-Centre
Coastal Pacific waters of Northern Kuril	June 2007	50–60	3.5	
Open Pacific	June 2007	50–60	2.5	

waters of Southern Kuril				
Southern Okhotsk Sea	June–July 1991	50–60	2.1–3.0	Starovoytov 2003
Okhotsk Sea	June–July 1993	30–40 40–50 50–60 > 60	2.4 2.8 2.1–2.2 1.3–1.7	Volkov et al. 1997
	June–July 1994	50–60 > 60	3.5 3.0	Starovoytov 2003
Southern Okhotsk Sea	October–November 2007	40–50	1.8	unpublished data of A.Ya. Efimkin, TINRO-Centre
Northern Okhotsk Sea	June–August 1991	50–60	2.9–3.9	Starovoytov 2003
	August 1997 September–October 1994 October 1998 September–October 1999 September–October 2000	> 50	2.4–4.3 2.1 2.2–2.6 2.5 2.4	Kuznetsova 2005
Okhotsk waters of western Kamchatka	August 1986	> 50	7.0	Gorbatenko and Chuchukalo 1989
Okhotsk waters of southwestern Kamchatka	June–July 1991	50–60	5.7	Starovoytov 2003
	June–July 1996	50–60 > 60	2.9–3.2 1.4	
Okhotsk waters of western Kamchatka	June–July 1996	50–60 > 60	2.7 2.8	
Northwest Pacific	February–March 1989	30–40 40–50 50–60	2.0–2.5 1.5–2.0 2.0–2.5	Chuchukalo 2006
	April–May 1989	30–40 40–50 50–60	4.5 3.0 3.5	Tutubalin and Chuchukalo 1992

Feeding of small immature chum on the feeding grounds in the North Pacific is not explored adequately yet. The data were collected only in 1989, 1991, and 2006, mostly in winter and spring seasons. Chum preys upon hyperiids, copepods, chaetognaths, salps, medusas, polychaets, pteropods, euphausiids, and fish juveniles in winter, when the most important groups and species are chaetognaths (*Sagitta elegans*, *Flaccisagitta enflata, F. lyra, Eukrohnia hamata),* salps (*Salpa aspera, S. fusiformis*, *Dolioletta gegenbauri*, *Doliolum denticulatum*), polychaets, and pteropods (*Carinaria cristata*, *C. galea*), because of sinking the crustaceans to deep layers of the sea for wintering (Tutubalin and Chuchukalo 1992; Chuchukalo 2006; Volkov et al. 2007). The feeding activity in winter is not high, daily rations don't exceed 2.5–3.0% and 2.0–2.5% for the fish with length 20–30 and 30–40 cm, correspondingly (Starovoytov 2003; Chuchukalo 2006). Chaetognaths become the dominant prey item in spring when their share in the chum ration reaches 72% for the fish with length 20–30 cm and 60% for the larger fish with length 30-40 cm (data for April-May, 1989); copepods and pteropods are also presented in the diet (Volkov et al., 2007). Spring rations are higher than winter ones in 1.5 times (Table 3). Because of lack of the data for summer season, the summer diet cannot be described. However, the chum feeding activity in early summer is estimated as high in compare with feeding in winter and spring (Tutubalin and Chuchukalo 1992).

Feeding area of small immature chum salmon expands to Bering Sea and adjacent waters in summer, where its diet was investigated in 1991–1995, 2003 and 2006. Euphausiids, hyperiids, pteropods, and young squids and fish are the base of the chum ration in the western Bering Sea (Volkov 1994; Volkov et al. 1997, 2007; Starovoytov 2003; Efimkin 2003, Efimkin et al. 2004; Chuchukalo 2006; Naydenko et al. 2007). The share of pteropods is the highest in the Commander Basin (Figure 13). The shares of euphausiids and nekton components increase here towards fall season that depends on seasonal changes in plankton and nekton. Fish larvae and juveniles (pollock, sand lance and other species) form the base of chum salmon diet in the eastern Bering Sea in autumn (data for 2003–2006), with exception of the Nunivak area where gelatinous zooplankton (mostly medusas) and appendicularians are also presented in the ration (Efimkin et al. 2004; Volkov et al. 2006; Volkov et al. 2007) (Figure 13). So, the ratio of fodder groups in the chum diet depends on the state of forage resource.

The summer diet of small (30–40 cm) immature chum salmon in the Pacific waters at eastern Kamchatka is formed mainly by euphausiids (up to 65%), copepods (30%, mainly *Eucalanus bungii*, to a lesser degree *N. plumchrus* and *N. cristatus*), hyperiids and fish juveniles (Volkov et al. 1997). Pteropods, euphausiids, chaetognaths, and gelatinous zooplankton make the chum diet in the Pacific waters of northern Kuril Islands in summer (data for 2004–2007), but pteropods only dominate of the south Kurils (data for 1991–1996) (Figure 19, 20).

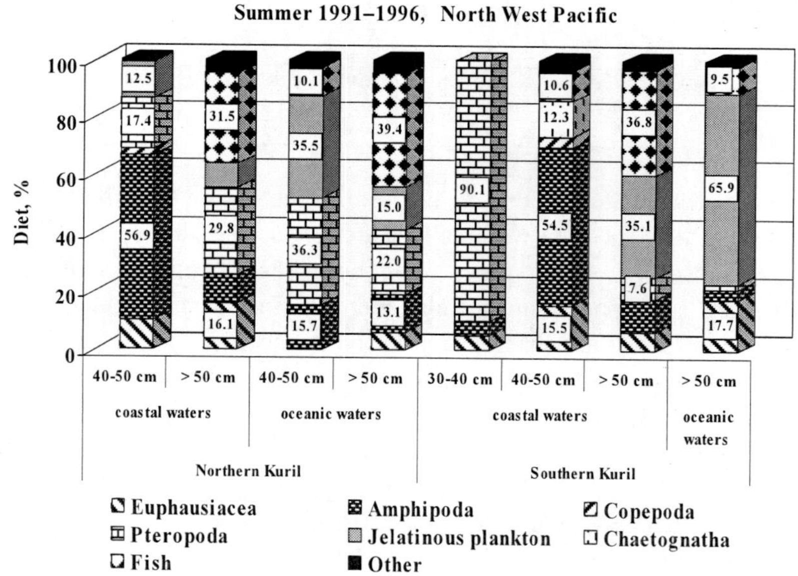

Figure 19. Diet (%) of chum salmon in the Pacific waters of the Kuril Islands in summer 1991–1996 (from Volkov et al. 1997; Starovoytov 2003; Naydenko and Kuznetsova 2002; Chuchukalo 2006; with modifications).

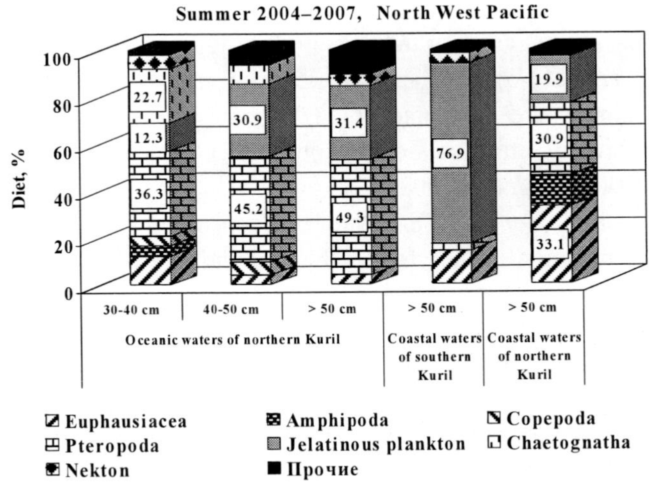

Figure 20. Diet (%) of chum salmon in the Pacific waters of the Kuril Islands in summer 2004–2007.

The immature chum salmon is not numerous in the Okhotsk Sea in summer-autumn. There is known that euphausiids, hyperiids, pteropods, and squid and fish juveniles dominate in chum diet in the southern part of the Sea, and hyperiids, euphausiids, decapod larvae in its northern part (Volkov et al. 1997; Chuchukalo 2006).

Feeding activity of small chums is high everywhere in summer and autumn; its daily rations are 1.6–4.6% of body weight in dependence on region (Table 3).

The diet of large-sized chum salmon includes the same items but in different ratio: the shares of nekton components, gelatinous zooplankton, and pteropods (in certain regions) are higher in the ration of large fish. For example, the highest shares of gelatinous zooplankton (mainly ctenophores, to a lesser extent salps and appendicularians) in chum salmon diet were observed in the Pacific waters of Kuril Islands (65.9–76.9%) in summer 1991–1996 and 2004–2007 and in the southern Okhotsk Sea (33.3–82.1%) in summer 1991–2003 and autumn 2003–2007 (Figure 19–21). The share of pteropods in diet was higher in the western Bering Sea and in particularly in the Pacific waters of Kamchatka and northern Kuril Islands (Figure 13, 19–20). In these regions pteropods biomass (especially *Clione limacina*) is usually highest in summer compared to other areas (Chuchukalo and Napazakov 1998). As distinct from small- and medium-sized chum salmon preying mainly on *L. helicina*, the large fish prefer another species of pteropods—*C. limacina*—which is why the highest share of pteropods in their ration is observed in summer, when *C. limacina* is the most abundant. The share of pteropods has significant year-to-year variability: for example, in the western Bering Sea it was 19.8–37.4% (*C. limacina* only) in the ration of adult chums (> 40 cm) in 1986 and 1990, but pteropods were absent in the ration of small and medium chum so far as *L. helicina* was not abundant in plankton; the chum of any size preyed upon both pteropoda species in autumn of 2003 as both were abundant; *L. helicina* only was abundant in 2006, that's its share was significant in the ration of small and medium fish. The high share of pteropods in the chum salmon diet is recognized by some researchers as "pteropod feeding specialization" (Volkov 1994; Radchenko 1994b; Volkov et al. 1997; Dulepova 1998; Starovoytov 2003; Efimkin 2003, Efimkin et al. 2004; Chuchukalo 2006). Among nekton components of the adult chum diet, young squids (in particular in the Aleutian Basin), juveniles of herring, pollock, sand lance, greenlings, and adults of Myctophidae species prevail. The highest share of young pollock is noted in the coastal areas of the western Bering Sea including the Karaginsky Bay, Anadyr Bay, and the area at Navarin Cape. The share of myctophids is the most prominent in the Commander Basin, where atka mackerel (*Pleurogrammus monopterygius*) and young perches are also important components of the diet in some years (Volkov 1994; Radchenko, 1994; Starovoytov 2002; Efimkin 2003; Volkov et al. 2006). The share of southern atka mackerel (*Pleurogrammus azonus*) in the southern deep-water areas of the Okhotsk Sea can reach 50% of the chum ration in the years of high abundance of the mackerel (Chuchukalo 2006).

The chum in the eastern Bering Sea has other feeding habits, which are determined by the structure of forage resources in this area (Pearcy 1988; Volkov et al. 2006; Volkov et al. 2007). High share of nekton in the adult chum diet is observed here, as well, but with walleye pollock domination. However, jelly fish dominate in the Nunivak area (Figure13). Medusas and hyperiids dominate in the chum diet in the North Pacific in winter, with prominent shares of euphausiids, copepods, and cephalopods (the data of 2006) and the shares of appendicularians and pteropods increasing in spring (Volkov et al. 2007). So, just as pink salmon, the chum salmon of any size adapt easily their feeding habits for conditions of certain feeding ground.

Feeding activity of adult chum is not high; daily rations in winter are usually 1.5–2.5% of body weight, with increasing to 3.0–4.5% in spring (Tutubalin and Chuchukalo 1992; Chuchukalo 2006), and to 3.8–4.6% in summer (for the feeding fish with length 30–50 cm). The activity decreases significantly for migrating fish during its migration to spawning grounds. So, the ration value varies considerably within the same region and season in dependence both on physiological state of fish and the state of forage resource (Table 3).

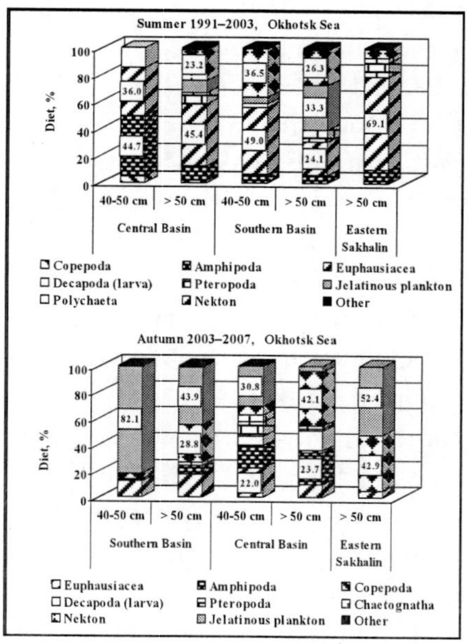

Figure 21. Diet (%) of chum salmon in the different regions of Okhotsk Sea in summer 1991–2003 and autumn 2003–2007 (from Volkov et al. 1997; Starovoytov 2003; Naydenko and Kuznetsova 2002; Chuchukalo 2006; with modifications and additions).

Sockeye salmon

Small-, medium- and large-sized sockeye have the same diet items, with changes of their ratio and size composition (Ito 1964; Burgner 1991; Chuchukalo et al. 1994a, 1995; Volkov 1994; Radchenko 1994a, b; Volkov et al. 1997, 2004, 2007; Efimkin et al. 2004). Generally, the role of hyperiids decreases with the fish age, and the role of euphausiids, pteropods, decapods, fish and squids increases.

Table 4. Daily diets of sockeye salmon

Region	Data	Length, cm	Daily diets, %	References
Western Bering Sea	June–July 1993	30–40 40–50 50–60	0.9 1.4 1.6	Volkov et al. 1997
Pacific waters at eastern Kamchatka	June–July 1993	30–40 40–50 50–60	2.2 0.5 1.0	Volkov et al. 1997
Western Bering Sea	June–July 1992	30–40 40–50 50–60	3.2 5.4 4.0	Chuchukalo 2006
	June–July 1993	30–40 40–50 50–60	2.9 3.1 2.1	
Pacific waters of eastern Kamchatka	June–July 1991	30–40 50–60	2.2 4.2	
	June–July 1993	30–40 40–50 50–60	2.4 3.1 2.2	
Okhotsk waters of western Kamchatka	June–July 1993	30–40 40–50 50–60	3.8 2.9 2.2	
Pacific waters of Northern Kuril Islands	June–July 1993	50–60	4.5	
Western Bering Sea	September–October 1992	30–40 40–50 50–60	3.2 5.4 4.0	Efimkin et al. 2004
	September–October 2002	30–40 40–50 50–60	2.8 1.8 1.8	
	September–October 2003	30–40 40–50 50–60	4.8 3.6 3.8	
	September–October 2004	30–40 40–50 50–60	3.2 2.8 1.0	Shuntov and Temnykh 2004
Northern Okhotsk Sea	September–October 1994	< 30	3.9	Kuznetsova 2005
	October 1998	< 30	2.2	
	September–October 2000	< 30	4.8	

As examples, the share of hyperiids was 10.7–10.9% in the ration of large-sized sockeye in western Aleutian and Commander Basins of Bering Sea in autumn of 2002–2007, that is considerably lower in compare with small- and medium-sized fish (46.1–52.5%) with corresponding increasing of the shares of euphausiids, pteropods, young fish, and decapods larvae; these differences were observed in summer of 1992–1993 and 2005, as well; in the eastern Bering Sea, larvae and juveniles of pollock, capelin, and sand lance dominate in the diet of sockeye with length 10–30 cm, but the larger fish consume mainly euphausiids (54.4% in the Bristol Bay), or decapods (51.9% at Nunivak Island) (Figure 8). Hyperiids dominate absolutely (77.6%) in the ration of small sockeye in the waters of eastern Kamchatka, but euphausiids (38.2%), cephalopods (13.6–43.1%) and gelatinous zooplankton (16.2–22.5%) are the main components of the adults ration (Volkov et al. 1997).

Crabs larvae, hyperiids, and copepods are the most important components of the medium-sized sockeye diet in the Okhotsk waters of western Kamchatka in summer, but fish and pteropods (in autumn) dominate here in the ration of large-sized fish (Chuchukalo 2006). The species composition of food can vary considerably from year to year, but *T. pacifica* is usually the most important among hyperiids, *Th. longipes* among euphausiids, *L. helicina* among pteropods, *N. cristatus* among copepods, and young *Gonatopsis borealis* and *Berryteuthis magister* among cephalopods (Volkov et al. 1997; Efimkin et al. 2004; Chuchukalo 2006).

Feeding activity of sockeye salmon varies even within the same season: the daily ration for medium-sized fish changes in summer from 0.9 to 3.8% of body weight, in autumn from 2.8 to 4.8%, and can reach 5.4% for large-sized fish. The lowest rations are observed in the final stage of anadromous migration when their values are 0.5–1.0% (Table 4).

Chinook salmon

Chinook salmon is predatory species, so nekton dominates in the diet of all its size groups, and large-sized plankton animals (mainly euphausiids) have less importance (Chuchukalo et al. 1994b; Volkov et al. 1997, 2004, 2007; Glebov 2000; Naydenko and Kuznetsova 2002; Efimkin et al. 2004; Chuchukalo 2006). Cephalopods (mainly *Gonatus kamtschaticus*, *G. borealis*) dominate in the chinook ration in the deep-water part of the western Bering Sea both in summer and autumn; their share was higher in the Aleutian Basin than in the Commander Basin (Figure 9). Squids make the base of the chinook diet in the Pacific waters of eastern Kamchatka, too. Fish prevail in the diet in the shelf areas of the western Bering Sea and in the eastern Bering Sea; in the Anadyr Bay and at Navarin Cape their share in the ration reaches 90–100% (Chuchukalo et al. 1994b; Glebov 2000; Efimkin 2003; Chuchukalo 2006; Kuznetsova 2006; Volkov et al. 2007). Among the fishes, myctophids (gen. Stenobrachius and Lampanyctus), capelin, atka mackerel, larvae and juveniles of pollock are the most important components in the western Bering Sea, and sand lance, herring, and bottom fishes prevail in the diet in the eastern Bering Sea. Megalops of crabs dominate in the chinook ration in the shallow Nunivak area (Volkov et al. 2007) (Figure 9). Chinook salmon prey mainly upon young pollock and herring in the northern Okhotsk Sea where the squids portion in the ration does not exceed 11–16%. The share of euphausiids increases in the central Okhotsk Sea. Japanese anchovy and sardine are exclusive food of chinook salmon in the southern Okhotsk Sea (Chuchukalo et al. 1994; Glebov 2000; Chuchukalo 2006). Squids (*Berryteuthis anonychus* with the length 12–20 mm), large-sized euphausiids (*Th. longipes* 12–22 mm), and fish form the chinook diet with length 40–50 cm in the North Pacific waters at Aleutian Islands in winter, but the squids become the only dominant group in spring (Volkov 2006; Volkov et al. 2007).

Table 5. Daily diets of chinook, coho and masu salmon

Region	Data	Length, cm	Daily diets, %	References
Chinook salmon				
Central Okhotsk Sea	July–August 1991 July–August 1993	< 40	4.6 4.4	Chuchukalo 2006
Northern Okhotsk Sea	October 1998	< 40	2.5	Kuznetsova 2005
Coho salmon				
Northern Okhotsk Sea	October 1998	50–70	3.1	Kuznetsova 2005
Central Okhotsk Sea	June–August 1991	50–70	3.0	Chuchukalo 2006
Okhotsk waters of western Kamchatka	June–August 1992	50–70	3.3	
Pacific waters of Kuril Islands	June–August 1991	50–70	3.4	
Pacific waters of Kuril Islands	June–August 1993	50–70	3.6	
Pacific waters of eastern Kamchatka	June–August 1993	50–70	2.1	
Northwest Pacific	April-May 1989	40–50 50–60	4.0 5.0	Tutubalin and Chuchukalo 1992
Masu salmon				
Tartar Strait (Japan Sea)	November 2003	30–40 40–50	2.9 3.3	unpublished data of A.E. Lazhentsev, TINRO-Centre
Primorye (Japan Sea)	November 2003	30–40	3.3	

Feeding activity of large chinook salmon is rather high in the summer-autumn period. For example, the daily ration of feeding chinook was 4.4–4.6% of body weight in the central Okhotsk Sea in July–August of 1991–1993. However, this activity decreases in the process of the fish migration to spawning grounds: the index of stomach filling becomes very low (the values 15–21% were registered in the waters of western Kamchatka in June–July of 1993). The feeding activity of feeding immature chinook in autumn is not high (daily ration 2.5%; Table 5), but in some cases very high stomach filling is observed (217 and 212$^{o}/_{ooo}$ in the western Bering Sea in November 1993 and December 1995) (Chuchukalo et al. 1994; Glebov 2000; Chuchukalo 2006). It is not high in winter and spring when the indices of stomach filling don't exceed 64–73% (Volkov 2006; Volkov et al. 2007).

Coho salmon

Coho has the same type of feeding, as chinook salmon: importance of nekton increases with the size of fish, accordingly with decreasing the role of planktonic components in the diet (Brouder 1987; Volkov et al. 1995, 1997, 2007; Chuchukalo 2006; Efimkin 2003; Efimkin et al. 2004). As examples: coho preyed mostly on squids (57.7–68.2% of the ration) and share of euphausiids was only 18.5%, with insignificant quantities of pteropods, mysids, hyperiids, and fish in the western Bering Sea in summer of 1991–1993 and 2005 (Volkov et al. 1997, 2007); the squids prevails in the diet in autumns of 2003–2007, as well, though the share of fish was rather high in the Commander Basin (Figure 10); fish (usually young pollock and herring) dominated together with crabs larvae in the diet of coho in the eastern Bering Sea in autumns of 2003–2005, and sand lance and euphausiids were the main components in 2006 (Volkov et al. 2007) (Figure 10); large-sized coho preyed upon greenling, anchovy, saury, sardine, juveniles of bottom fishes, and to a lesser degree upon euphausiids in the southern and central areas of the Okhotsk Sea and upon young pollock, northern smoothtongue, and other fishes in the area oft western Kamchatka in summer of 1991–1994 (Volkov et al. 1995; Volkov et al. 1997; Chuchukalo 2006).

However, in some cases the share of plankton was 100%, even in the diet of large fish. The coho feeding in winter season is not clearly understood, but it is known that the coho with length > 40 cm had high feeding activity and preyed upon squids and fish, in much less degree upon chaetognaths, euphausiids, and copepods in the open North Pacific in winter and spring of 1989 (Tutubalin and Chuchukalo 1992; Chuchukalo 2006) and upon squids (mainly *B. anonychus* with length 12–20 mm) and hyperiids (*T. pacifica* and *P. macropa*) in the Pacific waters at Aleutian Islands in winter of 2006 (Volkov 2006).

Feeding activity of coho salmon is rather high in summer and autumn. The indices of stomach filling changed in the range 86–300$^{0}/_{000}$ for the coho with length 50–70 cm in the Okhotsk Sea in summer of 1991–1993 and autumn of 1998, that corresponded to daily rations from 2.1 to 3.6% (Kuznetsova 2005). The high activity remained in winter and spring: in April-May of 1989 a daily ration of 4.0% was registered for fish with a length of 40–50 cm and a ration of 5.0% for fish with a length of 50–60 cm (Tutubalin and Chuchukalo 1992) (Table 5).

Masu salmon

Samples size on masu feeding habits (for individuals with FL > 30 cm) are sporadic. However, size of samples collected in the northwestern Japan Sea during November, 2003 is quite satisfactory. Fishes (anchovy, young greenling) form the base of the masu diet in the Japan Sea, and the joint share of planktonic components (hyperiids, euphausiids, copepods) doesn't exceed 4.0–6.1% of the food weight (unpublished data of A.E. Lazhentsev, TINRO). In the Okhotsk Sea, young pollock and capelin dominate in the diet of masu (30–50 cm) in the area of western Kamchatka, and young sand lance, capelin, saffron cod, and squids in the area at eastern Sakhalin (Gorbatenko and Chuchukalo 1989; Kuznetsova 2005). Masu preyed mainly upon fish (anchovy, sand lance, saury, mictophids, young pollock), gammarids (24.0–37.5% of food) and euphausiids (8.0–15.0%) in the Pacific waters of southern Kuril Islands in autumn-winter 2000 and 2001, while the role of squids was lower (2.0–2.5%) (Baginskiy 2002).

Masu salmon has medium feeding activity in autumn and winter, with daily rations changing from 2.9 to 3.3% of body weight (Table 5).

In general, feeding of salmon depends on both changing physiological state of fish and temporal and spatial changes of forage resorses. Since evolution adaptations of salmon are formed in the process of settling over the upper pelagic layer, their behavioral and morphological features are oriented toward utilization the huge reserves of this layer, in particular macroplankton and micronekton (Shuntov and Temnykh 2005). Therefore, on one hand, all salmon species have a high feeding plasticity and are able to change their feeding from one object to other in dependence on the state of forage resorses. The examples of different feeding habits of salmon between the western and eastern parts of Bering Sea, between the Aleutian and Commander Basins, between different parts of the Okhotsk Sea, between the Pacific and Okhotsk Sea waters on two sides of Kuril Islands chain, and others illustrate well this ability. On the other hand, salmon have definite selectivity relative to some fodder items, as to hyperiids and pteropods; this ability allows them to utilize more effectively forage resources in certain areas. At the same time, the efficiency of consumption and selectivity of salmon feeding are not determined only by morphological and physiological features of preys, but also by their abundance, in some extent.

Role of Pacific Salmon in the North Pacific Marine Ecosystem

The most important parameters for estimation the role of Pacific salmon in marine ecosystems, on the one hand, and for estimation the carrying capacity of pelagic layer to salmon, on the other hand, are the ratio between forage resources and nekton consumers and the volumes of food consumption by the whole nekton, and by Pacific salmon, in particular. Based on the summarized estimations presented by Shuntov (2001), Dulepova (2002), Shuntov and Temnykh (2007), and Volkov et al. (2007), the total stock of zooplankton in the epipelagic layer of the subarctic northwest Pacific, including the Okhotsk and Bering Seas, has decreased on 134.5 million tons in two decades (from 722.5 million tons in 1980 to 587.8 million tons in 2000), i.e., approximately in 20%. The decreasing is noted everywhere, from internal bays of the seas to deep-water open ocean. However, the large-sized fraction of zooplankton, the basis of feeding for nekton planktivores, had another changes: a slight decreasing from 1980 to early 1990s (from 560.8 to 500.5 million tons, or 8%) and a slight increasing in succeeding years (to 520.1 million tons). At that, the dynamics has some specific provincial features in certain regions. Generally, in spite of a definite fall of total abundance of macroplankton, its biomass is still rather high and is estimated recently, by regions, as 239.0 million tons in the Okhotsk Sea, 152.6 million tons in the Paific waters of Kuril Islands, 47.1 million tons in the Pacific waters of eastern Kamchatka, and 81.4 million tons in the western Bering Sea (Shuntov 2001; Dulepova 2002; Shuntov and Temnykh 2007). Following to Shuntov and Temnykh (2007), about 60% of this stock is distributed in the deep-water parts of the regions. That is important for correct estimation of forage reserves for salmon, whose feeding grounds are located mostly in deep-water areas where fish productivity is usually rather low, sometimes more than 10-fold lower (8–13-fold, by the regions) than in shelf and slope waters (Shuntov et al. 2007; Shuntov and Temnykh 2007).

Quantitative estimations of the resources of small pelagic fishes and squids, based on the same surveys data, show that this important for salmon forage reserve is also significant, in particular for shelf and slope areas (Nekton... 2003–2006; Shuntov and Temnykh 2007). The small pelagic nekton is able to compensate the lack of zooplankton forage resources, as in the eastern Bering Sea where larvae of fish and bottom invertebrates form the base of salmon diet in conditions of low abundance of large-sized zooplankton (Volkov et al. 2007).

The total resources of pelagic nekton had a prominent fall in late 1980s to the early 1990s when the walleye pollock stock had decreased and japanese sardine stopped its mass migration to suabarctic waters; later this lowering was partially compensated by rising the populations of herring, saury, japanese anchovy, capelin, several species of squids, and salmon themselves. Recently the resource of pelagic nekton is still lower than in 1980s but higher than in early 1990s (Shuntov et al. 1997; Shuntov and Temnykh 2004).

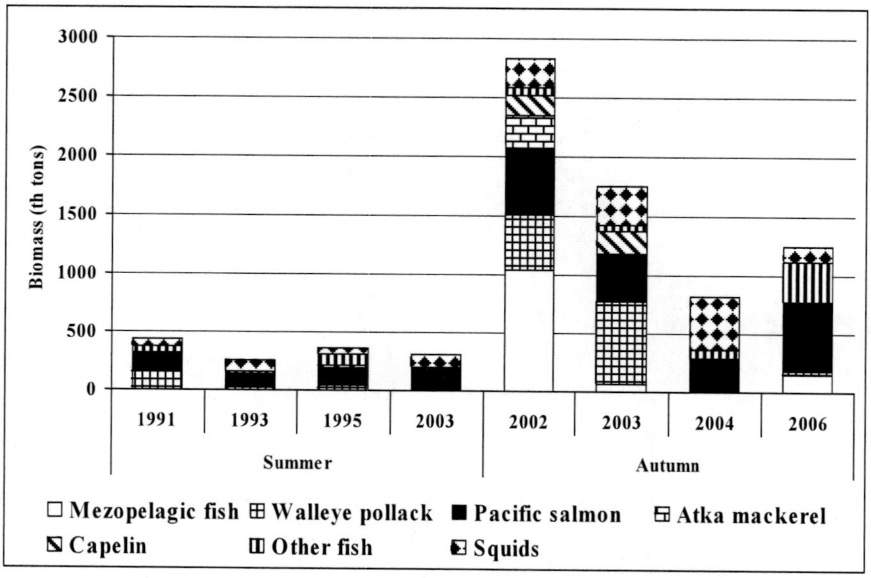

Figure 22. The biomass (thousand tons) of nekton in the upper epipelagic layer in the western Bering Sea during summer and autumn (from Shuntov et al. 1993b, 2007; Radchenko 1997; Temnykh et al. 2003, 2004; Shuntov and Temnykh 2004, 2005; Shuntov and Sviridov 2005; Temnykh 2005; 2006 Glebov et al. 2006; Starovoytov 2007; with modifications).

Changes in the abundance and structure of the main consumers affect the trophic structure of nekton community, as well. The volumes of consumption, energy fluxes direction, and trophic role of many mass species, including salmon, were changed.

First of all, it concerns the upper pelagic layer of the western Bering Sea, where the consumption of forage resources by nekton became considerably lower because of decreasing stock of walleye Pollock (Figure 22). It had consumed 5.5–6.6 million tons in autumn period in the years of its high abundance (Figure 23). These values of the pollock consumption in the years of its high abundance in the western Bering Sea are confirmed in other studies, as well (Efimkin and Radchenko 1991; Shuntov et al. 1993a, 2000; Radchenko 1994; Temnykh et al. 2004). The pollock distribution in the western Bering Sea is very variable and depends on its stock and dynamical environmental conditions (Stepanenko 1997, 2001; Nikolaev and Stepanenko 2006). The weakening of young pollock migration in the northwestern part of the Sea in 2004 and 2006 caused the prominent lowering of food consumption by this species: to 0.15 million tons. As the result, the total consumption by nekton (fishes and cephalopods) in 2004 and 2006 was in 3–5 times lower than in the years of high abundance of pollock and was estimated as 1.4–2.2 million tons (Figure 24). In opposite, the salmon biomass had 3-fold increasing since 1980s and their share in the total fish biomass in the upper pelagic layer of the western Bering Sea became 21–53% in 2002–2006 (Shuntov et al. 1993b, 2007; Temnykh et al. 2003; Shuntov and Temnykh 2004, 2005; Glebov et al. 2006; Starovoytov 2007). Therefore, the share of food consumption by salmon has increased from 4.4 to 58–70% of total consumption by fish (Figure 23), so they consume recently 52% of total food consumption by nekton (fishes and squids) in summer 2003 and 39–41% in autumn 2004 and 2006 (Figure 24). However, the salmon role in the trophic structure of the upper pelagic layer is still not so important as the pollock role in the years of its high abundance.

Grazing of certain groups of zooplankton changes considerably in dependence on fluctuations of both zooplankton and nekton populations. For example, pollock in the cases of its high abundance in the western Bering Sea (autumns of 2002 and 2003) consumed almost one-third of total stock of euphausiids: 3.5 and 3.1 million tons, but their grazing was estimated as 1/30 in the autumns of 2004 and 2006, when the pollock abundance was low (Naydenko 2007). The grazing on copepods at the same times had decreased in 10–20 times (Table 6), but decreasing of hyperiids grazing was not so drastic because of active consumption of this group by salmon, atka mackerel and other nekton species (Naydenko 2007). Total consumption of macroplankton by nekton in fall season had decreased from 1/11 in 2002–2003 to 1/40 in 2004 and 2006. The observed increase in plankton resources due to their lower consumption (considering intrinsically high values of plankton biomass) support good foraging environment Pacific salmon and other pelagic fish species.

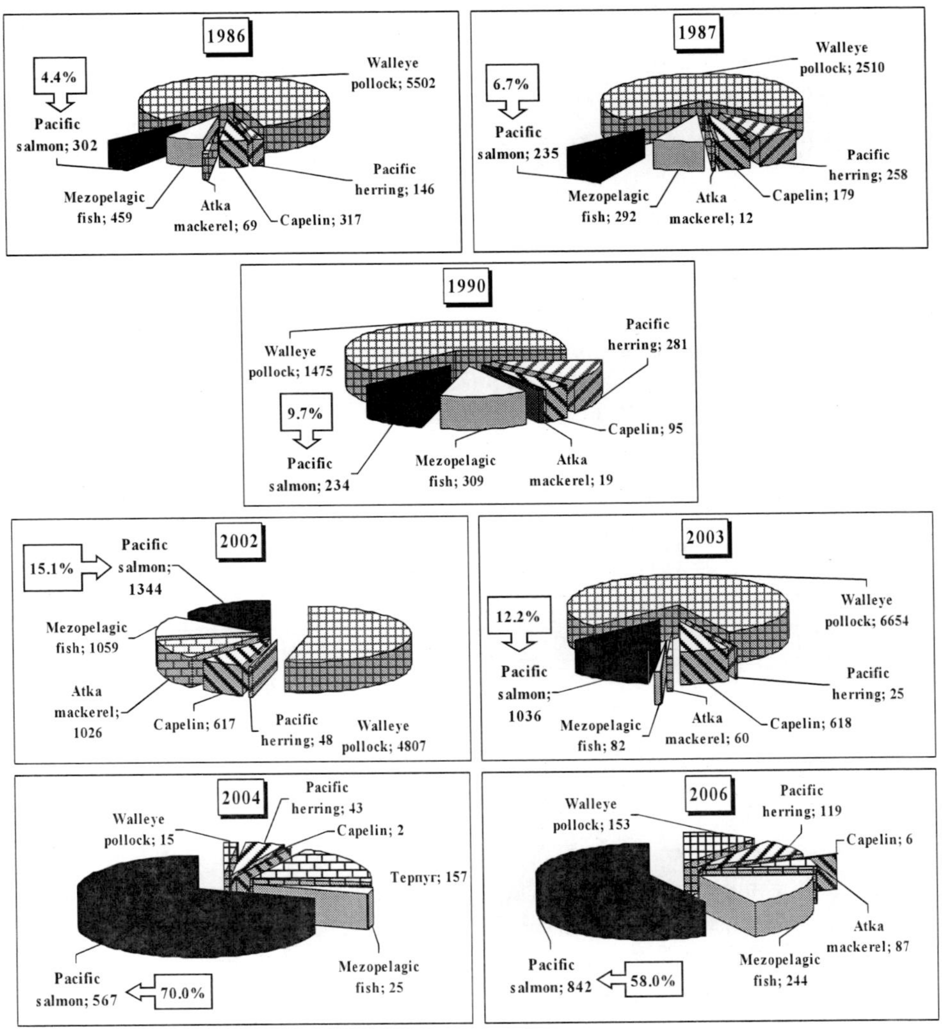

Figure 23. The consumption (thousand tons) of total forage resources by fish in the upper epipelagic layer in the western Bering Sea during autumn 1986–1987, 1990 and 2002–2006 (Naydenko 2007).

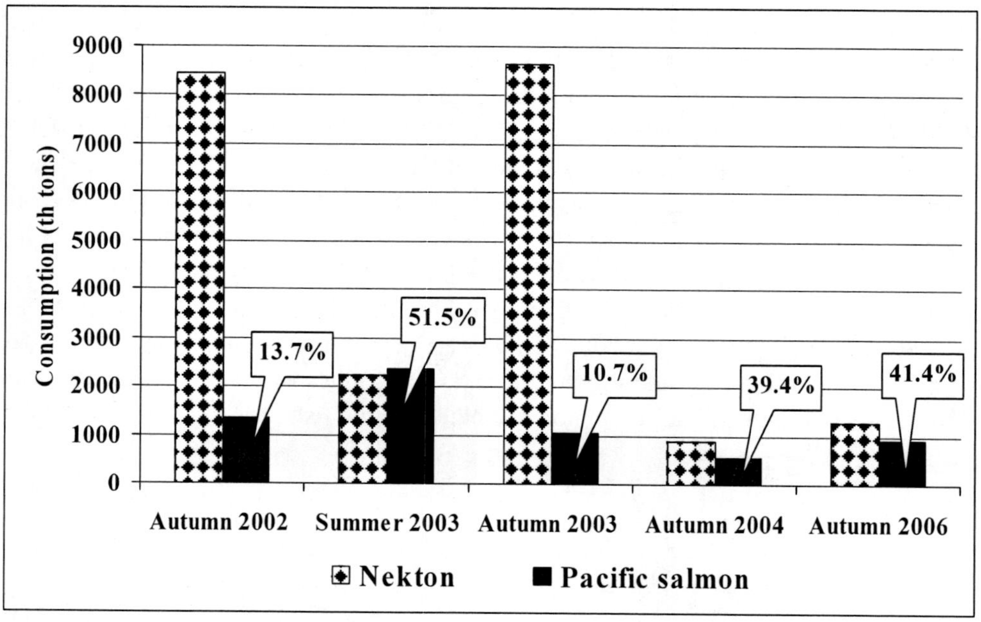

Figure 24. The consumption (thousand tons) of total forage resources by nekton and Pacific salmon in the upper epipelagic layer in the western Bering Sea during summer 2003 and autumn 2002–2006 (Naydenko 2007).

Table 6. The main zooplankton groups stocks (10^3 tons) in the upper pelagic layer of the western Bering Sea in autumn 2002–2006 and their grazing by nekton (numerator, consumption by the whole nekton; denominator, consumption by fish) (Naydenko 2007)

Years	Euphausiids		Copepods		Hyperiids	
	Biomass	Consumption	Biomass	Consumption	Biomass	Consumption
2002	9067	4082/3514	43722	2337/2333	2424	1047/1042
2003	10059	4141/3095	80438	2075/2071	1833	577/569
2004*	14429	709/222	10145	120/118	1796	294/293
2006	8568	1089/716	27035	226/224	1794	219/209

* only a part of the shelf was surveyed

Of course, only a few groups of zooplankton are actively consumed by nekton, but in the western Bering Sea the abundance of the groups preferred by nekton is sufficient. For example, summary biomass of only 5 the most important for nekton feeding species (*N. plumchrus*, *T. pacifica, C. limacina, L. helicina* and *Th. longipes*) was estimated here as 19.1 million tons in the autumn of 2004. Taking into account the higher production rate of zooplankton, a prominent difference between the reserves of macroplankton and food requirements of nekton could be seen.

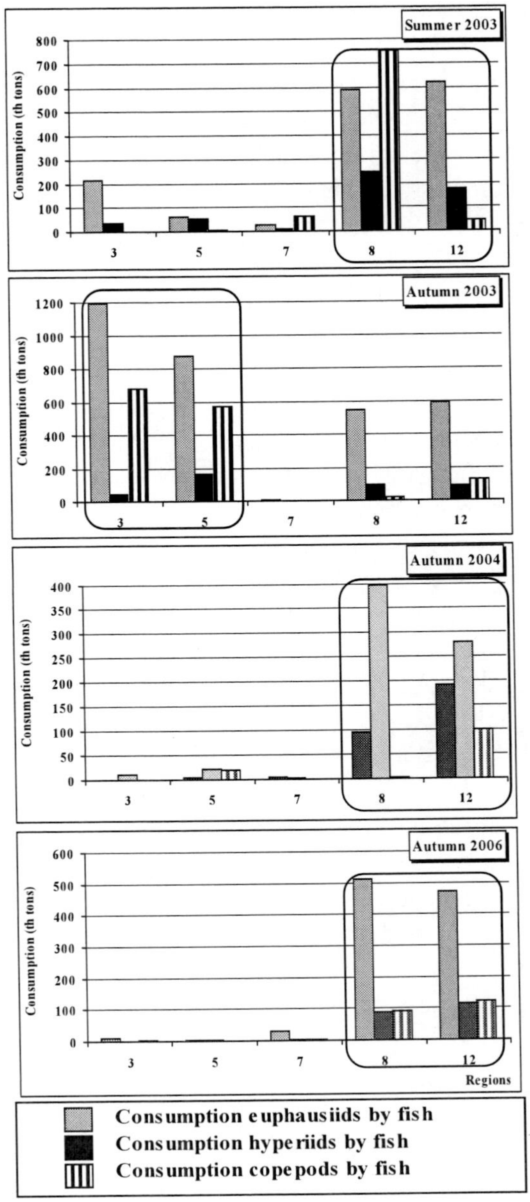

Figure 25. The consumption (thousand tons) of zooplankton forage groups by nekton in the upper epipelagic layer in the different regions of the western Bering Sea during summer 2003 and autumn 2003–2006 (Naydenko 2007) (names and number regions in Figure 1).

The satisfactory, at least, food supply for salmon in the western Bering Sea is proved by stable and high values of their daily rations (Table 2–5), against a background of considerable fluctuations of plankton and nekton abundance, and by permanent preference in their feeding for hyperiids, pteropods, and euphausiids.

Spatial dissociation of the main consumers and different time of their feeding are other important factors preventing the competition for food. For example, Pacific salmon, young atka mackerel, and squids fed in the deep-water areas of the western Bering Sea in summer 2003, and autumn of 2004, 2006 (Figure 25), but the main feeding grounds of pollock and capelin were located in shallow waters in the Anadyr Bay and at Cape Navarin where the consumption of zooplankton was very active in the autumn of 2003. Some competition for food is possible in cases of the feeding grounds overlapping, but it is mitigated by plasticity and electivity of salmon feeding, as well as by excess forage reserves.

Similar reconstruction of pelagic nekton community is noted for the Pacific waters of Kuril Islands. Total fish biomass fell here in early 1990s to 0.8 million tons that was in 5–8 times lower than in 1980s because of cessation of sardine migration and local pollock stocks depletion (Shuntov et al. 1993c, d, 1998; Ivanov 1998; Baitaluk et al. 2007). Accordingly, nekton consumption of the whole forage resources became 3 times lower (consumption of zoolankton 3.5 times lower) in 1990s than in 1980s (Ivanov 1998), in particular grazing of copepods was reduced, so far as this group formed the base of sardine and pollock diets in the region (Naydenko 2002). Perhaps, this deliverance of plankton resources became one of the reasons of predatory plankton bloom in the early 1990s together with intensification of alternative nekton species (anchovy, squids, saury and some others) migrations into these waters for feeding in summer-autumn (Dulepova 1996, 2002; Shuntov et al. 1993a, c, d, 1996; 1997; Shuntov and Dulepova 1996; Naydenko 2002). As the result, the fish biomass reached here 1.6–2.1 million tons in 2000s, and the total nekton biomass became 1.9–2.8 million tons (Figure 26). The value of total consumption of forage resources by nekton has increased here from 2.2 million tons in the summer of 1995 to 3.4–4.5 million tons in the summers of 2004, 2006, 2007 (Figure 27) (including the zooplankton grazing increasing from 1.6 to 2.9–3.7 million tons, correspondingly), and the value of salmon consumption has increased from 0.23 to 0.48–0.76 million tons (zooplankton grazing —from 0.18 to 0.41–0.71 million tons, correspondingly). However, even in the period of the highest salmon abundance in summer 2007, their share in the consumption of plankton by salmon did not exceed 24% (Naydenko 2008) (Figure 28). The salmon consume less quantity of forage resources than the species of lowboreal-subtropic complex or even mesopelagic fishes and squids in some years (Figure 27). Only in the short period of pre-anadromous migration beginning, before the mass arrival of south migrants, the salmon become the most important consumer of epipelagic nekton community (at least in cases of relatively low abundance of mesopelagic species).

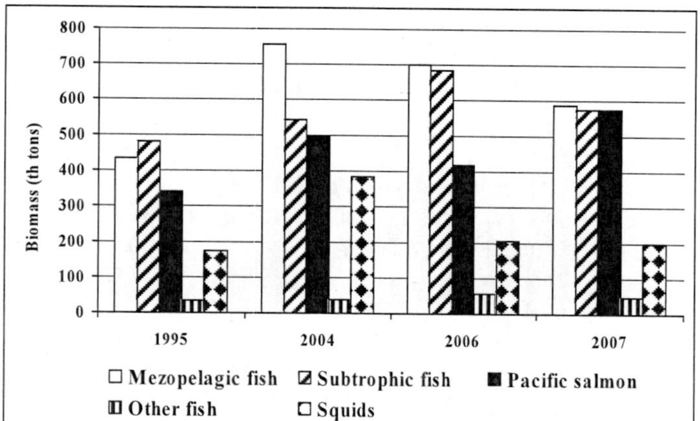

Figure 26. The biomass (thousand tons) of nekton in the upper epipelagic layer of Pacific waters of Kuril Islands in the summer (from Shuntov et al. 1998; Ivanov 1998; Baitaluk et al. 2007).

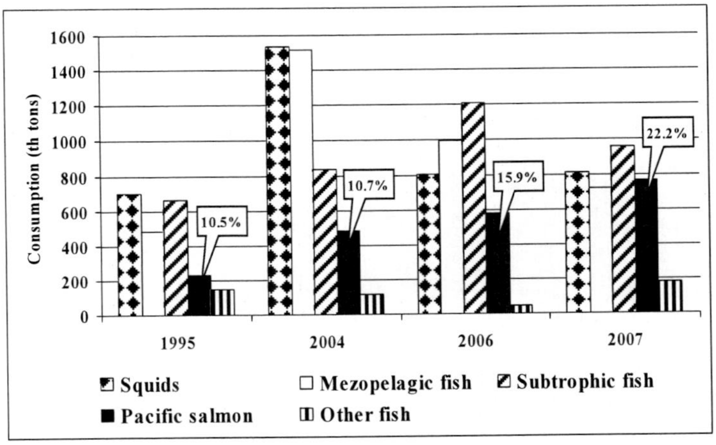

Figure 27. The consumption (thousand tons) of forage resources by nekton in the upper epipelagic layer of Pacific waters of Kuril Islands during summer (Naydenko 2008).

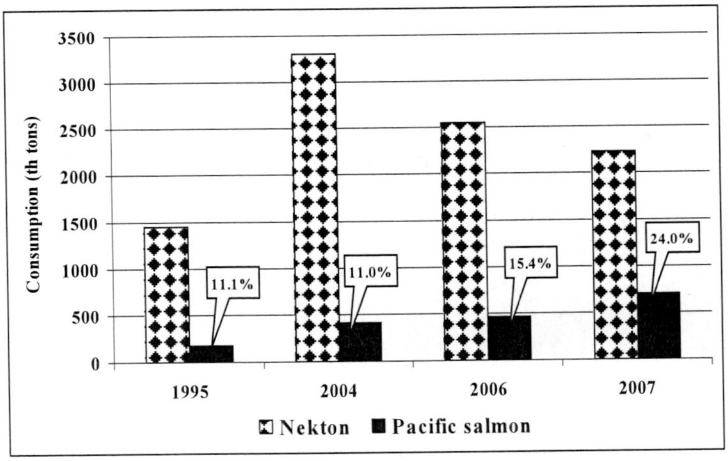

Figure 28. The consumption (thousand tons) of zooplankton forage organisms by nekton in the upper epipelagic layer of Pacific waters at Kuril Islands during summer (Naydenko 2008).

Mean stock of zooplankton in the Pacific waters of Kuril Islands is estimated as 167 million tons for summer-autumn period (from the data of 1996–2006) (Shuntov et al. 1993c,d; Shuntov 2001; Dulepova 2002; Shuntov and Temnykh 2007). The stock in summer was 64–100 million tons (by the data for of 1995, 2004, 2006, and 2007). Pacific salmon consumed only 0.2–0.8% of the summer stock (Figure 29). However, more than 80% of this value is the biomass of copepods and chaetognaths, which are minor items of the salmon diet. So, in the periods with low abundance of amphipods, pteropods, and euphausiids but high abundance of fish (for instance, in conditions of overlapping the pre-anadromous migration of salmon and feeding migration of subtropic species) the trophic relation could be strained in local areas. However, these periods cannot be long because of passing the migrating salmon through the region, so the competition cannot influence significantly on abundance or production parameters of the salmon. Quite satisfactory provision of salmon with the food in the area of Kuril Islands can be illustrated by the following example: a part of pink salmon migrating through this region in 1991 (when its abundance was very high) had low maturity coefficients and continued active feeding during migration; the feeding was successful so far as the diet contained the prey preferred for this species and the daily rations reached the value 8.8% of body mass, that is the evidence of sufficient forage reserves (Naydenko and Kuznetsova 2002).

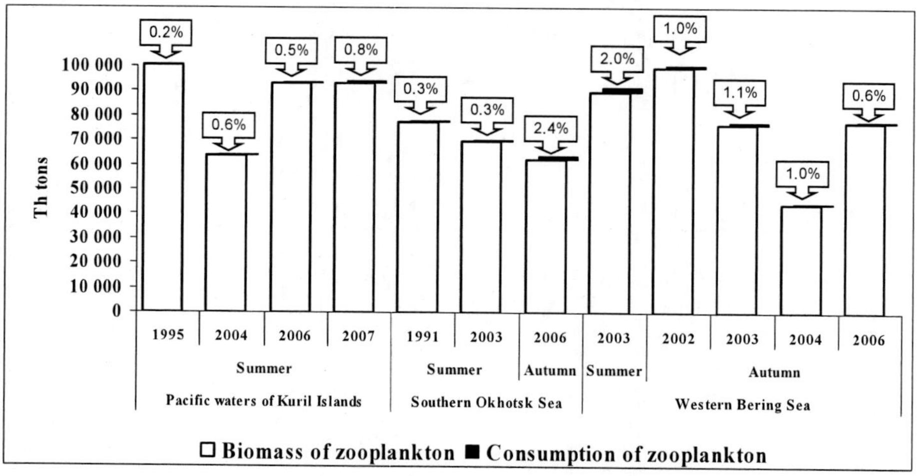

Figure 29. The biomass of zooplankton and its consumption (thousand tons) by Pacific salmon in the epipelagic layer of the Okhotks and Bering Seas and Pacific waters at Kuril Islands (from Naydenko 2007, 2008; with additions).

In the Okhotsk Sea, Pacific salmon are more abundant in its southern part where their anadromous and catadromous migrations pass in summer and autumn. The salmon biomass in this area is estimated as 0.24 million tons in summer (mainly adults of pink and chum salmon) and 0.12–0.49 million tons in autumn (mainly post-catadromous juveniles) (Figure 30) The salmon are rather stable component of summer-autumn nekton community of the southern Okhotsk Sea, while the abundance of other species is more variable that is the main factor of changing the role of salmon in the nekton community.

Total values of forage resources consumption by nekton in this area in summer change in the range from 2.3 (1991) to 4.7 (2003) million tons. In the first case, sardine and pollock were the main consumers (sardine consumed 1.0 million tons, pollock 0.8 million tons), and the share of salmon didn't exceed 13% (0.3 million tons). Note that the sardine was the main consumer in this region before 1991 (Ivanov 1998; Temnykh et al. 2004). In 2003, the main consumers were young south atka mackerel and capelin because of very high feeding activity (daily rations 8.5% and 4.1–18.6%, correspondingly): they consumed 73% of the whole nekton consumption (3.4 million tons). The squids grazed 18% (0.8 million tons), and the share of salmon was low again—6% (Figure 31).

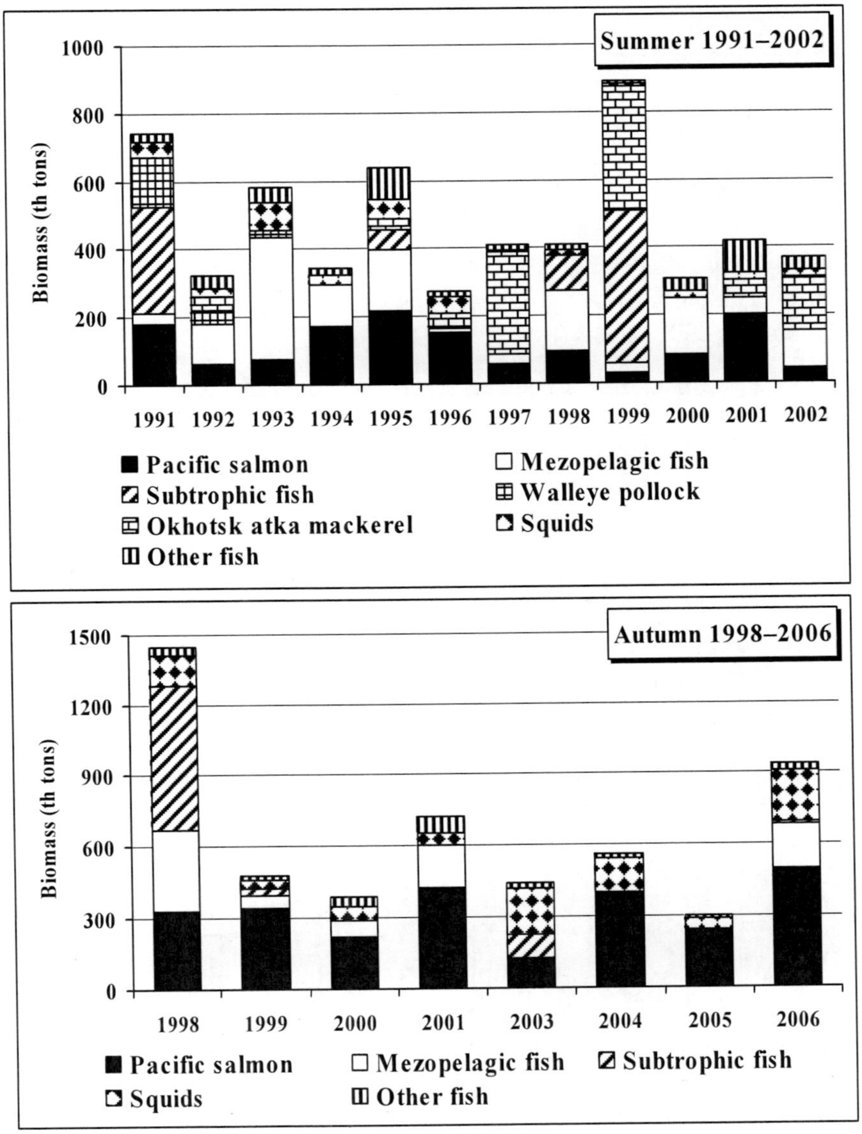

Figure 30. The biomass (thousand tons) of nekton in the upper epipelagic layer in the southern Okhotsk Sea during summer 1991–2002 and autumn 1998–2006 (from Shuntov 1989 a, b; Shuntov et al. 1993 c, d, 1998, Radchenko 1997; Radchenko et al. 1997; Temnykh et al. 2003, 2004; with modifications and additions).

The role of salmon increases in autumn when post-catadromous juveniles of pink and chum become more abundant, up to 95% of the nekton biomass (Figure 30). Their biomass here is higher in the western Bering Sea: in average 0.23 million tons of pink and 0.11 million tons of chum (Nekton… 2004, Temnykh 2004, Shuntov and Temnykh 2007). Other fishes, as japanese anchovy and smoothtongue, has a comparable biomass in some years only. In the autumn of 2006, the summary biomass of salmon in this area was 0.49 million tons (juveniles 0.44 million tons) and they consumed 1.5 million tons of food or 56% of the total consumption by epipelagic nekton (Figure 32). Similar estimations of food consumption by the pink and chum juveniles were made by Temnykh (2004): 1.3 and 0.4 million tons, correspondingly. The post-catadromous juveniles continue their feeding in the southern Okhotsk Sea till December-January (Shuntov 1989a,b; Radchenko et al. 1991; Starovoytov et al. 2004; Temnykh 2005), possibly because of high forage resources of zooplankton (Shuntov 2001; Dulepova 2002). In winter of 1992 the salmon juveniles were the most important consumers of zooplankton in the upper pelagic layer of this area (Figure 32). On the whole, salmon juveniles play more significant role in trophic relations of epipelagic nekton in the southern Okhotsk Sea than in the western Bering Sea: compare their shares of the total consumption by nekton 53% in the Okhotsk Sea (autumn 2006) and 1–8% in the western Bering Sea (autumns 2002–2006) (Figure 33).

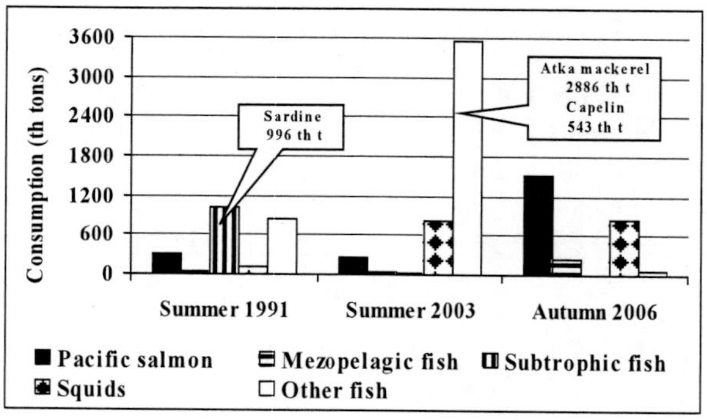

Figure 31. The consumption (thousand tons) of forage resources by nekton in the upper epipelagic layer in the southern Okhotsk Sea during summer 1991 and 2003 and autumn 2006.

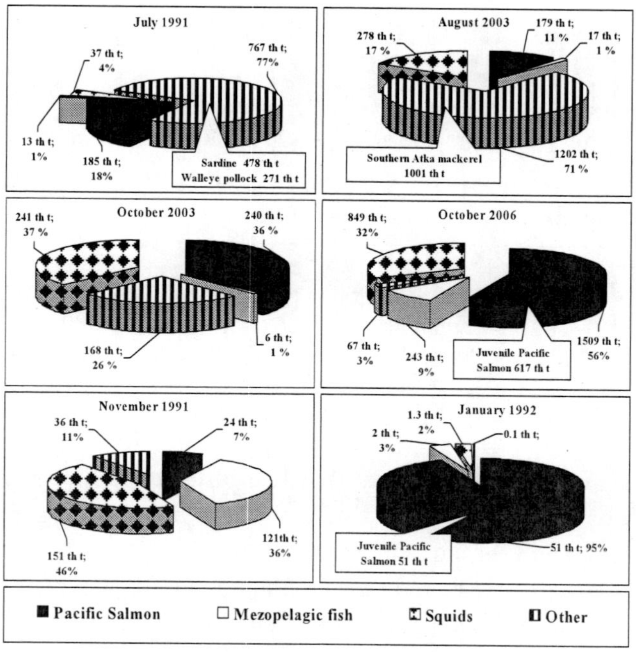

Figure 32. The month consumption (thousand tons) of total forage resources by nekton in the upper epipelagic layer in the southern Okhotsk Sea in different season.

Thus, the share of salmon in consumption of forage resources by epipelagic nekton is the highest in the western Bering Sea in summer and in the southern Okhotsk Sea in autumn, but the value of their consumption is not high in all investigated regions and changes from 0.2 to 2.4% of total zooplankton biomass (without production) in dependence on region and year (Figure 29). Besides zooplankton, small fishes and squids form a considerable part of the salmon ration. Their mean summary concentration is estimated by long-term trawl surveys as 0.1–3.9 g/m^2 in dependence on region (Shuntov and Temnykh 2007). Compare this with the mean concentrations of macroplankton animals preferable for salmon feeding, such as pteropods (1–2 g/m^2), hyperiids (2–8 g/m^2), and euphausiids (15–25 g/m^2). Obviously, the small nekton organisms are consumed actively by salmon in cases of temporary or local deficit of zooplankton food.

So far as salmon contribution in consumption of forage resources is insignificant, on the whole, the recent rise in their abundance would scarcely cause serious reconstructions in the trophic structure of the upper pelagic layer, in particular in the Far Eastern seas and northwest Pacific in summer and autumn. Indeed, any fluctuations in populations of the main consumers lead to changes in the trophic structure, but even multiple shifts in the biomass of highly-fluctuated species, such as pollock, herring, capelin, sardine, scomber, anchovy, and Japanese flying squid, result no more than fitting of trophic relations and ratios between energy fluxes, without any crisis situation in communities.

Let us consider two generalized schemes of trophic relationship in the epipelagic nekton community of the western Bering Sea for the autumns of 2003 and 2004 (Figures 34–35). On one hand, feeding orientation of different nekton groups is divergent: they consume different groups of plankton, which relaxes the competition for food. On the other hand, the main food relations of nekton were directed onto euphausiids and copepods in 2003, these two groups were grazed considerably by pollock that was abundant in that year, but most likely these energy fluxes switched to plankton predators (gelatinous zooplankton, chaetognaths, some copepods) in 2004 after nekton consumption decreasing. Unfortunately, the trophic relations within zooplankton community have not yet been well investigated.

Regarding the problem of carrying capacity of the upper pelagic layer for salmon, it is necessary to account for its dependence on many factors and high variability. From the point of view of food supply, it depends on the ratio between forage reserve and total abundance of its consumers. Relative to this, let us note once more that the recent rise of salmon abundance is backgrounded by a considerable (2–4-fold, on average) decrease in total nekton abundance in the upper pelagic layer, while zooplankton resources remain stable at a rather high level. So, the conditions are perfectly favorable for Pacific salmon feeding in the Far Eastern seas at present, and the salmon do not influence considerably the energy fluxes in the ecosystems in spite of rising abundance. The same conclusion was made earlier by Temnykh et al. (2004), Shuntov and Temnykh (2004) and Naydenko (2007, 2008).

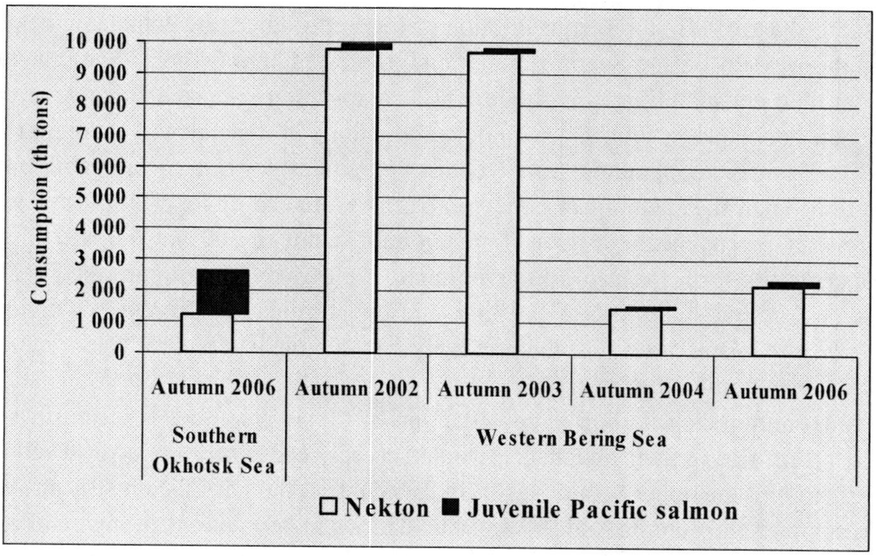

Figure 33. The consumption (thousand tons) of total forage resources by juvenile Pacific salmon and nekton in the upper epipelagic layer in the southern Okhotsk Sea during autumn 2006 and in the western Bering Sea during autumn 2002–2006 (Naydenko 2007; with additions).

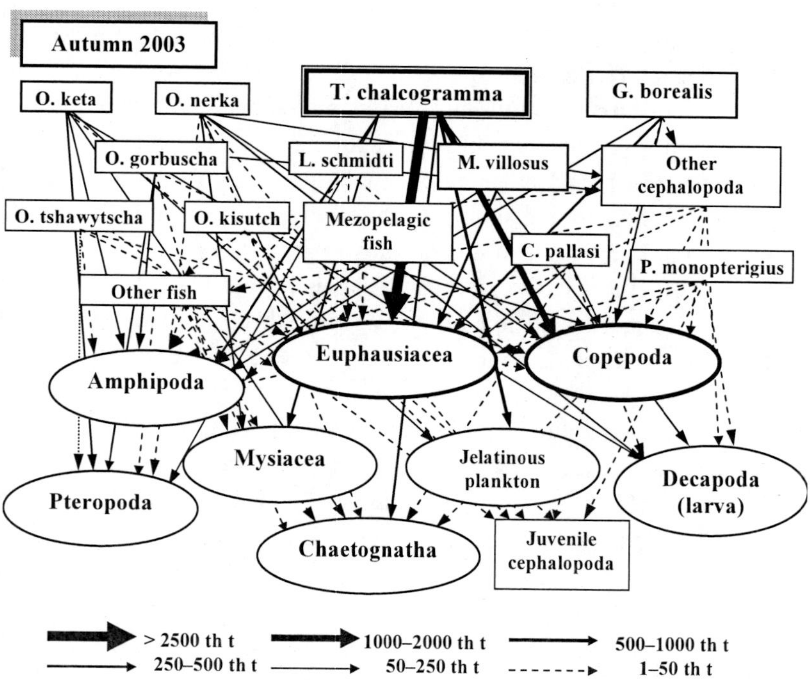

Figure 34. The scheme of the basic trophic relationships of most abundant nekton species and main zooplankton groups in the upper epipelagic zone of the western Bering Sea in autumn 2003 (thickness of arrows—volume of autumn consumption, in thousands of tons).

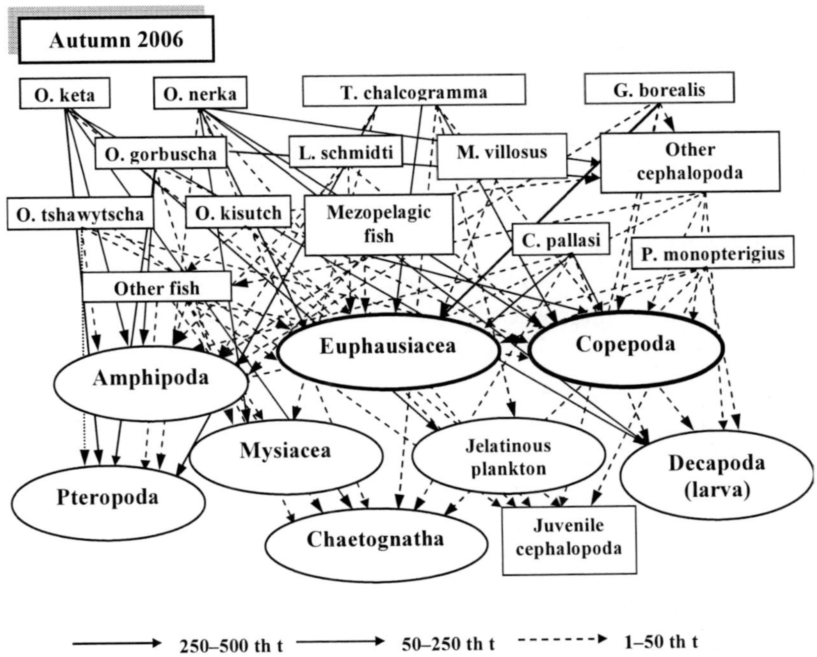

Figure 35. The scheme of the basic main trophic relationships of most abundant nekton species and main zooplankton groups in the upper epipelagic zone oa the western Bering Sea in autumn 2006 (thickness of arrows—volume of autumn consumption, in thousands of tons).

However, this conclusion is based on the data obtained in the Far Eastern Seas in summer and autumn only. Such calculations cannot be done for the open waters of the North Pacific in the winter and spring seasons because of the absence of representational data, which is an important gap in the knowledge on the carrying capacity of salmon. There are vague and contradictary opinions on the conditions of salmon feeding in the winter and spring. According to some researchers, zooplankton biomass in the North Pacific in winter is considerably lower compared to summer (Nagasawa 1999, 2000; Ishida et al. 2000; Fukuwaka et al. 2007), and feeding conditions of salmon are unsatisfactory, but others doubt these estimations and claim underestimation of zooplankton abundance because of methodological problems (Shuntov 2001; Shuntov et al. 1993). Detailed quantitative studies of both zooplankton state and salmon feeding in winter and spring are necessary, including daily rhythm of feeding activity, as well as studies of the feeding of other nekton species in their common winter feeding grounds. Other important problems of forage reserve estimation is the definition of energy value and biochemical composition of food, which is necessary for all basic assumptions of trophic relationships in communities and ecosystems. These problems, still unsolved, and vague and insufficient knowledge regarding the mechanisms of many processes and their interactions do not allow a realistic estimation of the carrying capacity of pelagic waters in their whole spatial-temporal ranges. To develop this experience, new surveys and new scientific approaches are necessary.

References

Andrievskaya, L. D. (1968). Feeding of juvenile salmon in the sea. Izv. *TINRO 64*, 73-80. (In Russian)

Andrievskaya, L. D. (1970). Feeding of juvenile salmon in the Sea of Okhotsk. Izv. *TINRO, 78*, 10-115. (In Russian)

Andrievskaya, L. D. (1975). The feeding of Pacific salmon in marine period of life. Abstract of Ph. D. thesis. *Petropavlovsk-Kamchatsky, 28*. (In Russian)

Azumaya, T. & Ishida, Y. (2000). Density interactions between pink salmon (*Oncorhynchus gorbuscha*) and chum salmon (*O. keta*) and their possible effects on distribution and growth in the Pacific Ocean and Bering Sea. *N. Pac. Anadr. Fish Comm. Bull. 2*, 165-174. (Available at http://www.npafc.org)

Baitaluk, A. A., Zavolokin, A. V. & Zavolokina, E. A. (2007). Pacific salmon in the nekton communities of the northwestern Pacific Ocean. Implementation of "Strategic plan for Far Eastern basin-scale Pacific salmon research program for 2006-2010 period." Vladivostok, TIRNO-Center: *Bull, 2*, 134-138. (In Russian)

Bigler, B. C., Welch, D. & Helle, J. H. (1996). A review of size trends among North Pacific salmon (Oncorhynchus spp.). *J. Fish. Aquat. Sci., 53(2)*, 455-465.

Brouder, R. D. & Pearcy, W. G. (1987). Diet feeding chronology, gastric evacuation and estimated daily ration of juvenile coho salmon, *Oncorhynchus kisutch* (Walbaum), in coastal marine environment. *J. Fish. Biol., 31(4)*, 465-77.

Bugaev, V. F., Welch, D. W., Selifonov, M. M., Grachev, L. E. & Eveson, J. P. (2001). Influence of the marine abundance of pink (*Oncorhynchus nerka*) on growth of Ozernaya River sockeye. Fish. *Oceanogr, 10(1)*, 26-32.

Burgner, R. L. (1991). Life History of Sockeye Salmon (*Oncorhynchus nerka) //* Pacific Salmon life history. C. Groot, & L. Margolis, (Ed). Canada: 1-117.

Chereshnev, I. A., Volobuev, V. V., Shestakov, A. V. & Frolov, S. V. (2002). Salmonoid fishes in Russian North-East. Dalnauka, *Vladivostok, 494*. (In Russian)

Chuchukalo, V. I., Volkov, A. F., Efimkin, A. Ya. & Kuznetsova, N. A. (1995). Feeding and daily ration of Sockeye Salmon (*Oncorhynchus nerka*) during the summer period. N. Pac. Anadr. *Fish Comm. Doc., 125*, 9 (Available at http://www.npafc.org)

Chuchukalo, V. I. (2006). Feeding and trophic relationship of nekton and nektobentos organisms in the Far Eastern Seas. Vladivostok, *TINRO-Centre, 484*. (In Russian)

Chuchukalo, V. I., Efimkin, A. Ya. & Kuznetsova, N. A. (1992[a]). Distribution of mezoplankton and feeding of fish in the upper epipelagic layer in the Western Kamchatka waters in summer 1991. *Vladivostok*, TINRO. Deposit in VNIERH № 1204-RH 92. 18. (In Russian)

Chuchukalo, V. I., Efimkin, A. Ya. & Kuznetsova, N. A. (1992b). Distribution of mezoplankton and feeding of fish in the upper epipelagic waters in the Sakhalin-Kuril region in summer 1991. Vladivostok, TINRO. Deposit in VNIERH № 1203-RH 92. *22.* (In Russian)

Chuchukalo, V. I., Efimkin, A. Ya. & Kuznetsova, N. A. (1992c). Distribution of mezoplankton and feeding of fish in the upper epipelagic waters of Bering Sea and East Kamchatka in summer 1991. Vladivostok, TINRO. Deposit in VNIERH № 1205-RH 92. *21.* (In Russian)

Chuchukalo, V. I. & Napazakov, V. V. (1998). The distribution and some features of the Pteropods biology in the Okhotsk and Bering Seas and adjacent waters of Pacific Ocean. Izv. *TINRO*, *124*, 584-601. (In Russian with English abstract)

Chuchukalo,V. I. & Kuznetsova, N. A. (2006). Feeding relations of fishes in epipelagial of the western Bering Sea in autumn 2003-2004. Izv. *TINRO*, *144*, 180-197. (In Russian with English abstract)

Chuchukalo, V. I., Volkov, A. F., Efimkin, A. Ya. & Blagoderov, A. I. (1994b). Distribution and feeding of Chinook salmon (*Oncorhynchus tschawytscha*) in the north-western Pacific Ocean. Izv. *TINRO*, *116*, 137-141. (In Russian with English abstract)

Chuchukalo, V. I., Volkov, A. F., Efimkin, A. Ya. & Kuznetsova, N. A. (1994a). Feeding habits and daily ration of Sockeye salmon (*Oncorhynchus nerka)* in summer. Izv. *TINRO*, *116*, 122-127. (In Russian with English abstract)

Davis, N. D., Fukuwaka, M., Armstrong, J. & Myers, K. (2004). Salmon food Habits studies in the Bering Sea, 1960 to Present. N. Pac. Anadr. Fish Comm. *Techn. Rep*., *6*, 24-28. (Available at http://www.npafc.org)

Dolganova, N. T. (2001). Composition, seasonal and interannual dynamics of plankton in the northwestern part of the Japan Sea. Izv. *TINRO*, *128(3)*, 810-889. (In Russian with English abstract)

Dulepova, E. P. (1996). Dynamics of structural characteristics in zooplankton communities of Sakhalin-Kuril region. *Biologiya moray*, *22(5)*, 295-299. (In Russian with English abstract)

Dulepova, E. P. (1998). The juvenile pink (*Oncorhynchus gorbuscha*) and chum (*O. keta*) salmon in the trophic structure of South part of Okhotsk Sea. Vopr. *Ichthiologii*, *38(5)*, 633-640. (In Russian with English abstract)

Dulepova, E. P. (2002). Comparative bioproductive of Far East Seas macroecosystems. Vladivostok, *TINRO-Centre*, *273*, (In Russian)

Efimkin, A. Ya. (2003). Salmon feeding in the western Bering Sea during summer and autumn 2002. Izv. *TINRO*, *134*, 120-134. (In Russian with English abstract)

Efimkin, A. Ya., Volkov, A. F. & Kuznetsova, N. A. (2004). Food habits of Pacific salmon in the Bering Sea during summer and autumn in 2003 and 2004. Izv. *TINRO*, *139*, 370-387. (In Russian with English abstract)

Efimkin, A. Ya. & Radchenko, V. I. (1991). Status of forage resource and distribution of epipelagic fishes in the western part of Bering Sea during the autumn period. *Biologiya moray*, *1*, 28-39. (In Russian with English abstract)

Glebov, I. A. (2000). Ecology of the chinook and coho salmon the Asian stocks during marine period of the life cycle. Abstract of Ph. D. thesis. Vladivostok: *TINRO-center*, *24*. (In Russian)

Glebov, I. A., Loboda, S. V., Zavolokin, A. V. & Zavolokina, E. A. (2006). Pacific salmon in nekton communities of the upper epipelagic layer in the western Bering Sea in August-October 2006. Implementation of «Strategic plan for Far Eastern basin-scale Pacific salmon research program for 2006–2010 period». Vladivostok, *TIRNO-Center*, *Bull*, *1*, 34-41. (In Russian)

Gorbatenko, K. M. (1996a). Feeding of juvenile pink and chum salmon in the Sea of Okhotsk epipelagic water layers in winter. Izv. *TINRO*, *119*, 234-243. (In Russian with English abstract)

Gorbatenko, K. M. (1996b). Seasonal features of zooplankton vertical distribution in the Okhotsk Sea. Izv. *TINRO, 119*, 88-119. (In Russian with English abstract)

Gorbatenko, K. M. & Chuchukalo, V. I. (1989). Feeding habits and daily ration of Pacific salmon (genus Oncorhynchus) in the Sea of Okhotsk in summer and autumn. Vopr. *Ichthiologii, 29(3)*, 456-465. (In Russian with English abstract)

Gritsenko, O. F. & Klovatsh, N. V. (2002). Self-regulation of Japan Chum Salmon Abundance. N. Pac. Anadr. Fish Comm. *Techn. Rep., 4*, 45-47. (Available at *http://www.npafc.org*)

Gritsenko, O. F., Klovatsh, N. V. & Bogdanov, M. A. (2001). New epoch of Pacific salmon existence // World Ocean: utilization of biological resources. Moscow: Inf.-*analytic collection, 2*, 168-174.

Heard, W. R. (1991). Life history of Pink Salmon (*Oncorhynchus gorbuscha*). Pacific salmon life histories. UDC Press. *Vancouver*, 121-30.

Ishida, Y., Ueno, Y., Nagasawa, K. & Shiomoto, A. (2000). Review of ocean salmon research by Japan from 1991 to 1998. N. Pac. Anadr. Fish Comm. *Bull, 2*, 191-201. (Available at http://www.npafc.org)

Ishida, Y., Ito, S., Kaeriyama, M., McKinnell, S. & Nagasawa, K. (1993). Resent changes in age and size of chum salmon (*Oncorhynchus keta*) in the North Pacific Ocean and possible causes. Can. *J. Fish. Aquat. Sci., 50*, 290-295.

Ito, J. (1964). Food and feeding habit of Pacific salmon (genus Oncorhynchus) in their oceanic life. Bull Hokkaodo Reg. Fish. *Res. Lab, 29*, 85-97.

Ivanov, O. A. (1998). Epipelagic community of fishes and squids in the Kuril Islands area of the Pacific Ocean in 1986-1995. Izv. *TINRO, 124*, 3-54. (In Russian with English abstract)

Kaeriyama, M., Nakamura, M., Edpalina, R. & Bower, J. R. et al. (2004). Changes in feeding ecology and trophic dynamics of Pacific salmon (Oncorhynchus spp.) in the central Gulf of Alaska. Fish. *Oceanogr, 13(3)*, 197-207.

Kaeriyama, M. (1996). Changes in bogy size and age at maturity of a chum salmon *Oncorhynchus keta*, Population Released from Hokkaido in Japan. N. Pac. Anadr. *Fish Comm. Doc., 208, 9*, (Available at http://www.npafc.org)

Karpenko, V. I. (1998). Early marine period of the life cycle of Pacific salmon. *VNIRO*, Moscow, *165*, (In Russian)

Klovach, N. V. (2003). Ecological consequences of large-scale cultivation of chum salmon. *VNIRO*, Moscow, *164*. (In Russian)

Klovatch, N. V. (2000). Tissue Degeneration in Chum Salmon and Carrying Capacity of North Pacific Ocean. N. Pac. Anadr. *Fish Comm. Bull., 2*, 83-88. (Available at http://www.npafc.org)

Kogan, A. V. (1963). About daily ration and feeding rhythm of ziege of Tsimlyanskiy reservoir. *J. Zool, 42*, 596-601. (In Russian with English abstract)

Kosenok, N. S. & Naydenko, S. V. (2008). Features of feeding and daily rations of Pacific salmon (*Oncorhynchus keta*, Walbaum) in the western Bering Sea in summer-autumn period. *Biologiya moray, 34(1)*, 20-29. (In Russian with English abstract)

Koval, M. V., Erokhin, V. C. & Tepnin, O. B. (2007). Features of forage base formation, feeding, and biological indices of pacific salmon juveniles in the East Kamchatka waters in the falls of 1998-2005. Izv. *TINRO, 151*, 423-449. (In Russian with English abstract)

Kuznetsova, N. A. (2005). Feeding and trophic relationship of nekton in the upper epipelagic layer of the Northern of the Sea of Okhotsk. Vladivostok, *TINRO-Centre*, *236*. (In Russian)

Kuznetsova, N. A. (2006). Feeding of Pacific salmon in the eastern Bering Sea in 2005. N. Pac. Anadr. *Fish Comm. Doc.*, *987*, *9*. (Available at http://www.npafc.org)

Kuznetsova, N. A., Farley, E. V., Moss, J. H. & Cieciel, K. (2007). Feeding of Pacific salmon in the eastern Bering Sea in August-October 2007. Implementation of "Strategic plan for Far Eastern basin-scale Pacific salmon research program for 2006-2010 period." Vladivostok, *TIRNO-Center*, *Bull*, *2*, 81-86. (In Russian)

Lazhentsev, A. E. & Bokhan, L. N. (2001). Feeding of juvenile chum and pink salmon in the Okhotsk Sea in autumn. Izv. *TINRO*, *128*, 791-798. (In Russian with English abstract)

Maksimov, V. V. & Smorodin, V. P. (2004). Feeding and trophic relationship of juvenile salmon (Salmonidae) in the southwestern Bering Sea in autumn. Researches of water bioresources of Kamchatka and a northwest part of Pacific Ocean. *Petropavlovsk-Kamchatskij*, *KoTINRO*, 233-245. (In Russian)

Maximenkov, V. V. & Maximenkova, T. V. (2004). Food relationships of young salmon and other fishes in the eastern part of the Sea of Okhotsk. Izv. *TINRO*, *139*, 340-348. (In Russian with English abstract)

Mikodina, E. V., Minkova, N. O., Klovatsh, N. V. & Laptesha, T. I., et al. (2002). Biochemical structure of bodies and muscles of Okhotsk Sea chum salmon in connection with the phenomenon of flabbiness of muscles. *J. Tr. VNIRO*, *141*, 94-103. (In Russian with English abstract)

Nagasawa, K. (2000). Winter zooplankton biomass in the Subarctic North Pacific, with a discussion on the overwintering survival strategy of Pacific salmon (Oncorhynchus spp.) N. Pac. Anadr. *Fish Comm. Bull.*, *2*, 21-32. (Available at http://www.npafc.org)

Naydenko, S. V., Efimkin, A. Ya., Kuznetsova, N. A. & Kosenok, N. S. (2005). Food habits and tropic position of Pacific salmon in the Bering Sea epipelagic communities in autumn 2000-2004. N. Pac. Anadr. *Fish Comm. Doc.*, *876*, 30. (Available at http://www.npafc.org)

Naydenko, S. V. (2002). Trophic structure of nekton in the epipelagial of the South Kuril region in summer, 1991-1995. Izv. *TINRO*, *130(2)*, 618-652. (In Russian with English abstract)

Naydenko, S. V. (2007). Implication of Pacific salmon in trophic structure in upper pelagic layer in the western Bering Sea in summer and autumn of 2002-2006. Izv. *TINRO*, *151*, 214-239. (In Russian with English abstract)

Naydenko, S. V. (2008). Implication of Pacific salmon in trophic structure in upper pelagic layer in the pacific waters of Kuril Islands in summer. S.M. Konovalov Memorial Meeting: Current state of water bioresources. Vladivostok, *TINRO-Centre*, 393-395.

Naydenko, S. V. & Kuznetsova, N. A. (2002). Trophic relations of Pacific salmon in the epipelagical zone of the South Kuril region. Izv. *TINRO*, *130(2)*, 570-594. (In Russian with English abstract)

Naydenko, S. V., Efimkin, A. Ya. & Lazhentsev, A. E. (2007). Regional diversity of juvenile pink salmon diet in autumn in the Bering, Okhotsk and Japan Seas. N. Pac. Anadr. *Fish Comm. Bull*, *4*, 117-126. (Available at http://www.npafc.org)

Nekton of the north-western Pacific Ocean. Abundance, biomass and species ratio. 2005. V. P. Shuntov, & L. N. Bocharov, (Edited) by Vladivostok, *TINRO-Centre*, *544*. (In Russian with English abstract)

Nekton of the north-western part of the Japan Sea. Abundance, biomass and species ratio. 2004. V. P. Shuntov, & L. N. Bocharov, (Edited) by Vladivostok, *TINRO-Centre*, *226*. (In Russian with English abstract)

Nekton of the Sea of Okhotsk. Abundance, biomass and species ratio. 2003. V. P. Shuntov, & L. N. Bocharov, Vladivostok, (Edited) by *TINRO-Centre*, *643*. (In Russian with English abstract)

Nekton of the western part of the Bering Sea. Abundance, biomass and species ratio. 2006. V. P. Shuntov, & L. N. Bocharov, (Edited) by Vladivostok, *TINRO-Centre*, *416*. (In Russian with English abstract)

Nemchinova, I. A. & Kovalenko, S. A. (2004). Feeding of fry chum (*Oncorhynchus keta*) in summer season in the southern part of the Sea of Okhotsk. Izv. *TINRO*, *138*, 311-319. (In Russian with English abstract)

Nikolaev, A. V. & Stepanenko, M. A. (2006). Stock assessment and spatial distribution of the eastern Bering Sea pollock (*Theragra chalcogramma*) in summer 2004 г. Izv. *TINRO*, *145*, 168-190. (In Russian with English abstract)

Novikova, N. S. (1949). About possibility of daily ration estimate at nature conditions. *Vestnik MGU*, *9*, 107-111. (In Russian)

Pearcy, W. G., Brodeur, R. D. & Shenker, J. M., et al. (1988). Food habits of Pacific Salmon and Steelhead Trout, midwater trawl catches and oceanographic conditions in the Gulf of Alaska, 1980-1985. Bull. Ocean. *Res. Inst.*, *25(2)*, 29-78.

Prahova, N. V., Kulikov, M. Yu. & Ulyanov, M. Yu. (1980). Diet composition of South Kuril pink salmon in the southern part of feeding grounds and in the waters of Iturup Island. Izv. *TINRO*, *104*, 113-115. (In Russian with English abstract)

Radchenko, V. I. (1994a). The Sockeye Salmon *Oncorhynchus nerka* in the Bering Sea — ecological description of species. Izv. *TINRO*, *116*, 42-59. (In Russian with English abstract)

Radchenko, V. I. (1994b). Composition, structure and dynamics of nekton communities in the epipelagic zone of the Bering Sea. Abstract of Ph. D. thesis. *Vladivostok*, *24*. (In Russian)

Radchenko, V. I., Volkov, A. F. & Figurkin, A. L. (1991). About winter feeding of pink salmon in the Sea of Okhotsk. *Biologiya moray*, *6*, 88-90. (In Russian with English abstract)

Radchenko, V. I., Melnikov, I. V. & Volkov, A. F., et al. (1997). Conditions and environment, structure of plankton and nekton communities of the epipelagic zone of the southern part of the Sea of Okhotsk and adjacent waters in summer 1994. *Biologiya moray*, *23(1)*, 15-25. (In Russian with English abstract)

Ruggerone, G. T., Zimmermann, M., Myers, K. W., Nielsen, J. L. & Rogers, D. E. (2003). Competition between Asian pink salmon (*Oncorhynchus gorbuscha*) and Alaskan sockeye salmon (*O. nerka*) in the North Pacific Ocean. Fish. *Oceanogr*, *12(3)*, 209-219.

Salo, E. O. (1991). Life history of Chum Salmon (*Oncorhynchus keta*). Pacific salmon life histories. UDC Press. *Vancouver*, 233-309.

Shiomoto, A., Tadokoro, K., Nagasawa, K. & Ishida, Y. (1997). Trophic relations in the subarctic North Pacific ecosystem: possible feeding effect from pink salmon. *Marine Ecolog. Progr. Ser.*, *150*, 75-85.

Shuntov, V. P. (2001). Biology of the Russian Far Eastern Seas. Vladivostok, *TINRO-Centre Vol., 1*, 580. (In Russian)

Shuntov, V. P. (2005). Strategic plan for Far Eastern basin-scale Pacific salmon research program for 2006-2010 period. *Vladivostok, TIRNO-Center, 62*. (In Russian)

Shuntov, V. P. & Dulepova, E. P. (1996). Current status and interannual dynamics of the demersal and pelagic communities in the ecosystem of Okhotsk Sea. Izv. *TINRO, 119*, 3-32. (In Russian with English abstract)

Shuntov, V. P., Volkov, A. F. & Efimkin, A. Ya. (1988[a]). Composition and modern state of pelagic fish communities in the western Bering Sea. *Biologiya moray, 2*, 56-65. (In Russian)

Shuntov, V. P., Volkov, A. F. & Efimkin, A. Ya. (1988b). Composition and modern state of pelagic fish communities in the Pacific Ocean off Kamchatka and Kuril Islands. *Biologiya moray, 4*, 54-62. (In Russian with English abstract)

Shuntov, V. P., Volkov, A. F. & Matveev, V. I. (1986). Peculiarities of formation of productive zones in the Sea of Okhotsk in autumn. *Biologiya moray, 4*, 57-65. (In Russian with English abstract)

Shuntov, V. P., Volkov, A. F., Temnykh, O. S. & Dulepova, E. P. (1993a). Walleye pollock in ecosystems of Russian Far-Eastern Seas. Vladivostok, *TINRO, 426*. (In Russian)

Shuntov, V. P. & Temnykh, O. S. (2004). Is the North Pacific carring capacity over in connection with salmons high abundance: myths and reality. Izv. *TINRO, 138*, 19-36. (In Russian with English abstract)

Shuntov, V. P. & Temnykh, O. S. (2005). New views on ecology of Pacific salmon during the marine period of life. V.Ya. Levanidov Biennial Memorial Meeting. *Vladivostok, Dalnauka, Vol. 3*, 594-610. (In Russian with English abstract)

Shuntov, V. P. & Temnykh, O. S. (2007). Total estimation of forage resources (macroplankton and small nekton) for Pacific salmon in the Far East Seas and adjacent ocean waters of the north-western Pacific Ocean. Implementation of "Strategic plan for Far Eastern basin-scale Pacific salmon research program for 2006-2010 period." *Vladivostok, TIRNO-Center: Bull, 2*, 260-266. (In Russian)

Shuntov, V. P., Volvenko, I. V., Volkov, A. F., Gorbatenko, K. M., Shershenkov, S. Yu. & Starovoytov, A. N. (1998). New data about condition of pelagic ecosystems of the Okhotsk and Japan Seas. Izv. *TINRO, 124*, 139-177. (In Russian with English abstract)

Shuntov, V. P., Temnykh, O. S. & Glebov, I. I. (2007). Some aspects of international program BASIS (2002-2006) implementation by Russia. Izv. *TINRO, 151*, 3-34. (In Russian with English abstract)

Shuntov, V. P., Radchenko, V. I., Dulepova, E. P. & Temnykh, O. S. (1997). Biological resources of the Russian Far Eastern Economic Zone: structure of pelagic and benthic communities, contemporary status, long-term dynamics trends. Izv. *TINRO, 122*, 3-15. (In Russian with English abstract)

Shuntov, V. P., Radchenko, V. I. & Chuchukalo, V. I., et al. (1993c). Composition of plankton and nekton communities of the upper epipelagic zone of the Sakhalin-Kurils region during anadromous migration of salmon. *Biologiya Morya, 4*, 32-43. (In Russian with English abstract)

Shuntov, V. P., Radchenko, V. I., Lapko, V. V. & Poltev, Yu. N. (1993d). Distribution of Pacific salmon in the waters of the Sakhalin-Kurils region during the period of anadromous migration. Vopr. *Ichthiologii, 33(3)*, 348-358. (In Russian with English abstract)

Shuntov, V. P., Radchenko, V. I., Lapko, V. V. & Poltev, Yu. N. (1993b). Distribution of Pacific salmon in the western Bering Sea and adjacent waters of the north-western Pacific Ocean in the period of anadromous migration. Vopr. *Ichthiologii, 33(3)*, 337-347. (In Russian with English abstract)

Shuntov, V. P., Lapko, V. V., Balanov, A. A. & Startsev, A. V. (1995). Interannual changes in anadromous migration of salmon in the Sakhalin-Kurils region. *Biologiya Morya, 21(2)*, 116-124. (In Russian with English abstract)

Slabinsky, A. M. (2007). Vertical structure of plankton community in the Bering Sea in summers of 2003-2004. Izv. *TINRO, 151*, 403-416. (In Russian with English abstract)

Starovoytov, A. N. (2003). Chum salmon (*Oncorhynchus keta* (Walbaum)) in the Far East Seas—biological description of the species. 2. Diet composition and trophic linkages of chum salmon in the Far East Seas and adjacent waters of the north-western Pacific Ocean. Izv. *TINRO, 133*, 3-34. (In Russian with English abstract)

Starovoytov, A. N. (2007). Trends in abundance and feeding of chum salmon in the western Bering Sea. N. Pac. Anadr. Fish Comm. *Bull, 4*, 45-51. (Available at http://www.npafc.org)

Starovoytov, A. N., Glebov, I. I. & Sviridov, V. V. (2004). Current status of nekton epipelagic communities in the western Bering Sea. Vopr. Ichthiologii 5 (1/17), 6-27. (In Russian with English abstract)

Stepanenko, M. A. (1997). Interannual variability of spatial differentiation of pollock *Theragra chalcogramma* in the north-western Bering Sea. Vopr. *Ichthiologii, 1*, 19-26. (In Russian with English abstract)

Stepanenko, M. A. (2001). The state of stock, interannual variability of recruitment, and fisheries of the Eastern Bering Sea pollock in 1980-1990s. Izv. *TINRO, 128*, 145-152. (In Russian with English abstract)

Temnykh, O. S. (2004). The Asian pink salmon during the marine period of life: biology, spatial differentiation, status and role in the pelagic communities. Abstract of Dc. Sc. thesis. Vladivostok: *TINRO-Centre, 48*. (In Russian)

Temnykh, O. S. (2005). The international program BASIS: the first results. *J. Ryb. Knoz, 3*, 21-24. (In Russian)

Temnykh, O. S., Starovoytov, A. N., Glebov, I. I., Merzlyakov, A. Yu. & Sviridov, V. V. (2003). Pacific salmon in the epipelagic fish communities of the south Okhotsk Sea. Izv. *TINRO, 132*, 122-153. (In Russian with English abstract)

Temnykh, O. S., Glebov, I. I., Naydenko, S. V., Starovoytov, A. N. & Efimkin, A. Ya. (2004). Contemporary status of Pacific salmon in the communities of the Far Eastern Seas. Izv. *TINRO, 137*, 28-44. (In Russian with English abstract)

Tutubalin, B. G. & Chuchukalo, V. I. (1992). Feeding of Pacific salmon (genus Oncorhynchus) in the northern Pacific Ocean in winter and spring. Biological resources of Pacific Ocean. *VNIRO*, Moscow, 77-85. (In Russian)

Volkov, A. F. (1994). Features of pink, chum, and sockeye salmon feeding habits during the anadromous migration. Izv. *TINRO, 116*, 128-136. (In Russian with English abstract)

Volkov, A. F. (1996a). Method of zooplankton sampling. Izv. *TINRO, 119*, 306-311. (In Russian with English abstract)

Volkov, A. F. (1996b). Pacific salmon fry feeding in the Sea of Okhotsk in fall and winter seasons. *Okeanologya, 36(1)*, 80-85. (In Russian with English abstract)

Volkov, A. F. (2006). Plankton and salmon feeding in the North Pacific in winter 2006 (the survey of RV Kaiyo-Maru). Izv. *TINRO, 147*, 265-275. (In Russian with English abstract)

Volkov, A. F., Efimkin, A. Ya. & Kuznetsova, N. A. (2007a). Plankton communities in the Bering Sea and some areas of the North Pacific in 2002-2006. Izv. *TINRO, 151*, 338-364. (In Russian with English abstract)

Volkov, A. F., Efimkin, A. Ya. & Kuznetsova, N. A. (2007b). Results of the BASIS studies on Pacific salmons feeding habits in 2002-2006. Izv. *TINRO, 151*, 365-402. (In Russian with English abstract)

Volkov, A. F., Efimkin, A. Ya., Kuznetsova, N. A. & Slabinsky, A. M. (2004). Description of the Bering Sea plankton population in the autumn of 2003 (The results of the BASIS joint Russian-Japanese-U.S. Expedition). Izv. *TINRO, 139*, 282-307. (In Russian with English abstract)

Volkov, A. F., Efimkin, A. Ya. & Chuchukalo, V. I. (1997). Regional feeding habits of asian salmons in summer period. Izv. *TINRO, 122*, 324-342. (In Russian with English abstract)

Volkov, A. F. & Chuchukalo, V. I. (1986). Manual for the Study of Fish Food Habits. *TINRO*, Vladivostok. 31.

Volkov, A. F., Kuznetsova, N. A. & Slabinsky, A. M. (2005). Current structure of planktonic community in the Bering Sea during the summer-autumn 2002-2004 (The research conducted by the BASIS Joint Russian-Japanese –U.S. Expedition). Izv. *TINRO, 143*, 219-239. (In Russian with English abstract)

Volkov, A. F., Kosenok, N. S. & Kuznetsova, N. A. (2006). Results of Pacific salmons feeding behavior studies in frames BASIS Program in 2000-2004. Izv. *TINRO, 144*, 198-218. (In Russian with English abstract)

Volkov, A. F., Chuchukalo, V. I., Efimkin, A. Ya. & Glebov, I. I. (1995). Feeding of coho salmon (*Oncorhynchus kisutch*, Walbaum) in the Sea of Okhotsk and north-western Pacific Ocean. Vopr. *Ichthiologii, 35(6)*, 840-843. (In Russian with English abstract)

Watanabe, K. (2000). Adults returning and ages at return of Chum Salmon around Nortern Japan. N. Pac. Anadr. *Fish Comm. Doc., 490*, 5. (Available at http://www.npafc.org)

Yatsu, A. & Kaeriyama, M. (2005). Linkages between coastal and open-ocean habitats and dynamics of Japanese sardine. *Deep-Sea Research II, 52*, 727-737.

Yurovitskiy, Yu. G. (1962). About feeding of blue bream (*Abramis ballerus*) of Rybinskiy reservoir. Vopr. *Ichthiologii, 2(2)*, 350-360. (In Russian with English abstract)

Zavarina, L. O. & Shevlakov, E. A. (2004). Possible mechanism of forming chum salmon cyclic abundant generations in the north-east coast of Kamchatka. Conversation of biodiversity of Kamchatka and coastal waters: Materials of 5th scientific conference. *Petropavlovsk-Kamchatsky, KamchatNIRO*, 52-55. (In Russian)

Zavolokin, A. V., Efinkin, A. Ya., Slabinsky, A. M. & Kosenok, N. S. (2007). Prey selectivity and food supply of some common fish species from the upper epipelagic layer in the western Bering Sea and Pacific areas. *Vestnik SVNTS DVO RAN, 3*, 33-49. (In Russian with English abstract)

grounds in the area they are staying with their animals in summer. It is also said that fishermen in the fjords has had preference to fishermen working on trawlers due to larger uncertainty in income. Each household can be allocated three fishing grounds, while seniors can in general only have one. If the senior has connections to the primary sector he can have three licenses.

Co-Management Arrangements in the Industry

Management of the wild Atlantic salmon is the principal responsibility of the Ministry of Environment, and the Directorate for Nature Management is the executing authority on several issues. The salmon management regime is characterized by cooperation between different sectors and interested parties. The bill on National Salmon Rivers and Salmon Fjords passed through the Norwegian Parliament in 2007, making a selection of water systems and fjords to undergo particular measures for protection of the wild Atlantic salmon. A special cooperative body was established, with representatives from the Directorate of Fisheries, The Norwegian Water Resources and Energy Directorate, the Norwegian Food Safety Authority and the Directorate for Nature Management. The interests of owners, fishermen and other parties are organized through particular associations or groups, like river owners, landowners, sports fishermen associations, professional fishermen organizations and environmentalist organizations. The Directorate for Nature Management takes on the task of arranging consultative meetings where stakeholders meet with the Directorate as well as with other representatives of the authorities involved in salmon management.

The indigenous population in Finnmark, the Sámi, has a seat in these consultative meetings. So also, does FeFo. In addition, the Sámi has a particular channel for partaking in management; the arrangement of consultations. The consultation procedure is developed based on the ILO 169 convention, which says that governments shall "consult the peoples concerned, through appropriate procedures and in particular through their representative institutions, whenever consideration is being given to legislative or administrative measures which may affect them directly" (article 6.1.)[3]. The procedure is limited to consultations between the state government and the Sámi Parliament, but in certain situations it can, according to article 6, implicate consultations with local Sámi stakeholders. The interests of the Sámi traditionally living by the sea can thus be promoted through several channels; through special organizations' participation on consultative meetings, represented by fishermen's organizations or land and river owner associations, through the Sámi Parliament's representation in the consultative management body, but more sovereign through the consultation procedure.

In recent years several regional professional management bodies has been established, and the County Governor are obliged to consider their advice when taking measures and developing new regulations. Guidelines for the management of salmon are now given for 5-year periods, and the current guidelines for the period 2008-2012 asks for strengthening of local participation in management.

[3] http://www.ilo.org/public/english/standards/egalite/itpp/convention/manual.pdf

It should also be noted that both local and private management are already substantive. In addition to the governmental institutions' work on improving reproduction and survival of Atlantic salmon, private interests have contributed with substantial resourses, first and foremost by measures improving the living conditions in areas for fry and the construction of stairs in rivers to improve spawning opportunities.

CATCH AND PARTICIPATION

The 170 tons of salmon caught in Finnmark in 2006 were caught in 1150 fishing localities. The distribution of localities between private owners and FeFo in the same year is presented in Figure 1. The figure also illustrates the distribution of localities between the many municipalities in Finnmark. Private ownership dominates in cities like Alta and Hammerfest, while the opposite is the case in smaller communities like Gamvik, Lebesby and Sør-Varanger. The distribution between private owners and FeFo is very similar to the situation ten years earlier when the state was the major owner of property.

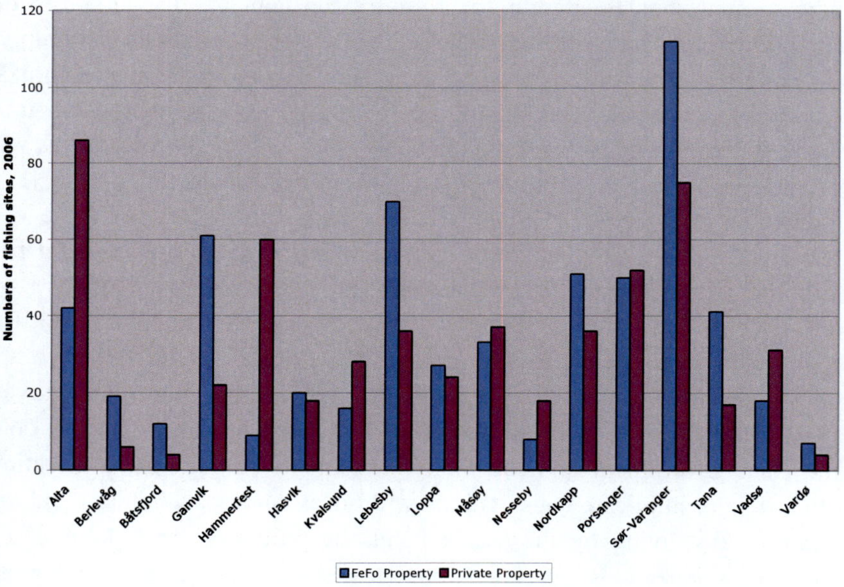

Figure 1. Number of salmon fishing sites in use, 2006, Finnmark.

The total catch has varied during the last decade. While catches totalled about 250 tons in the year 2000, they were only half the size four years later. Recently the catches are up again. Figure 2 shows catches in three selected years for all municipalities in Finnmark. The selected series are not long enough to make statements about trends, but some features can be highlighted. As all municipalities but two are represented in the figure, it is fair to say that the sea fisheries are familiar to the inhabitants of this county. Compared to the county's cod fishery, which is the mainstay of fishermen, with catches reaching 50 000 tons, catches of salmon are however, low. But the salmon fisheries provide a summer activity in which many fishermen participate. The figure also illustrates the uncertainty regarding size on catch from one year to another. The very solid figures on total catches from the year 2000 were representative for almost all municipalities in Finnmark. Still, Tana and Berlevåg stand out with very large catches in this year, much larger than in the years before and ahead.

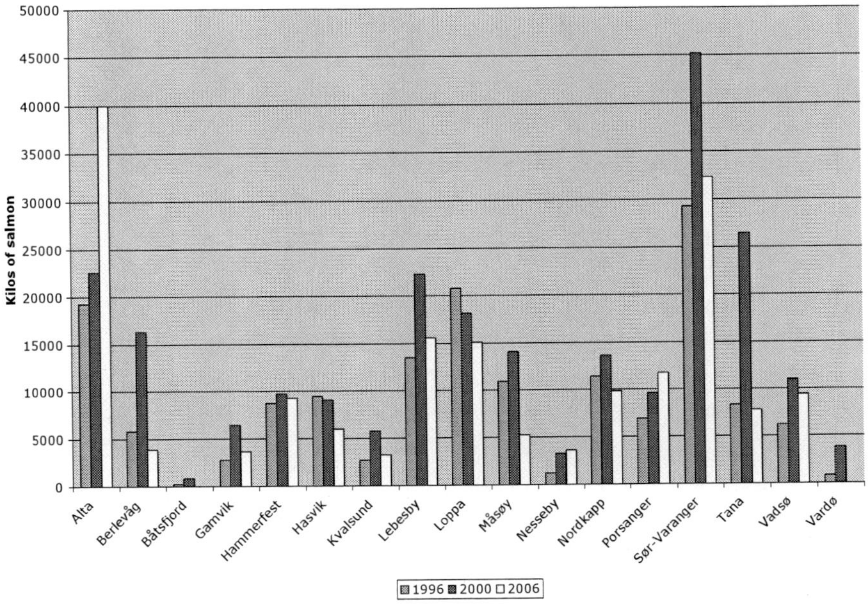

Figure 2. Salmon catches on municipality level. Finnmark 1996, 2000, 2006.

Catches in 2006 are generally higher than for the year 1996 in most municipalities. Rise in effort is an explanation easily available for explaining the larger catch. This is not the case in the salmon fishery. The number of localities used for fishing has in fact been reduced from 1996 till 2006, both on grounds owned by FeFo and on private property. Looking at the last ten years the reduction is somewhat larger on property managed by FeFo than private property. The reduction amounts to 700 localities in ten years, a reduction of close to 40%. The catches are 12% up in the same period. Larger stocks of salmon in the same period explain this. A higher sea temperature is thought to be a main contributor to the larger growth. But management efforts contribute as well. The growth in catches in the municipality of Alta is particularly large. The Alta River is a water system regulated for the purpose of hydroelectric power, and efforts have been made to improve the spawning for salmon and survival of fry, through new minimum levels of water in the river. The amount of gear in the Alta fjord has lessened, as in other fjords, but the catch in the localities still in use is more than doubled.

A fairly large number of vacant localities exist. In 2006 only 596 out of 1600 localities managed by FeFo were in use. Localities close the communities are attractive, while those on weather beaten grounds and in remote areas are less popular and stay vacant. There are several explanations to reductions in number of localities in use. Of particular importance is the reduction of qualified applicants due to closing of local farms and reduced number of fishermen. Not many younger people participate in the salmon fishery, and the fishermen's average age is increasing.

The Cultural Importance of Sea-Based Salmon Fisheries

Salmon has been part of Norwegian coastal culture for hundreds of years. It is an important carrier of traditions and it is visible throughout Norway's cultural history—in old rock carvings, myths and folklore, religion, poetry, painting, handicraft and language. Several places along the coastline carry the Norwegian name on salmon, "laks," like Laksefjorden and Lakseberget. Both Norwegian and Sámi language contains words describing salmon at different stages of its lifecycle.

History also tells stories about domestic servants with a clause in their contracts that salmon should not be served more than three times a week. These well-known stories reveal the former importance of salmon in economy and households. The economic value is far less today, and hardly anyone can claim to have wild salmon served three times a week. Availability is in fact so poor that many have never eaten wild salmon. The market for wild salmon has faded, partly because the professional buyers are gone, making sale and distribution difficult, and partly because competition from farmed salmon has pressed the price on wild Atlantic salmon, making earnings much harder.

The economic importance of wild salmon fisheries is poorly documented, but NOU (1999: 9) makes a calculation on revenues. The calculations are based on figures from 1990s when 3000 fishermen caught about 500 tons a year. Each fisherman made an average catch of 200 kilo. With a wholesale price of about 40 NOK per kilo, the average income to each fisherman was 8000 NOK. When costs to maintain a double bag net was about 6000 NOK per year the sum left for the fisherman was poor. It is necessary to point out that these are average figures, and some fishermen obviously earn more, some of them considerably more. A more important point to stress, however, is that the fishery for Atlantic salmon carries other qualities than the size of economic income. Fishing for salmon provides immaterial rewards to the fisherman. The activity, as well as the money, is motivating to its participants.

New revenues from the salmon fishery are discussed. While the river-based salmon fisheries have provided income through the sale of licenses for a long time, this is new in the sea-based fishery. Improved infrastructure in the fishery have raised interest and given some experience on sale of fishing rights to the tourism sector, and FeFo have signaled interest in developing the sale of sea fishing licenses, as a product to the tourism sector.

The Future of the Sea-Based Fishery for Atlantic Salmon

In a parliamentary proposition (St.prp.nr. 32 [2006–2007]), the government proposed the need for stronger regulations and management measures in the fishery for Atlantic salmon.

The view advocated is that the sustainability of the particular salmon stocks that are harvested must be documented. In the rivers the salmon stocks are often clearly defined. Often the river holds only one stock native to that particular river. Thus, the particular river stocks can be investigated and the health statuses documented one by one. According to a Norwegian classificatory system for salmon stocks, around half of the 450 salmon rivers hold stocks in need of special attention (Anon., 2007). As a catch of salmon at sea will often hold salmon from several stocks, it is likely that some are below conservation limits. Thus, the new management steps implicate, as stated by the proposition, "the sea-based fishery for Atlantic salmon must be curbed, and probably phased out in certain areas" (St.prp.nr. 32 [2006–2007], p. 35).

The Ministry of Environment also suggests phasing out the use of bend nets entirely. Except from Finnmark, this gear has been prohibited in all counties since 2002. The background to this suggestion is that this gear is more selective, targeting medium size females in particular. The fisheries management body has voiced concern over size and gender structures in the salmon stock, noting that the number of medium size salmon is less than in historic figures. Adding to this is the problem relating to damage to the fish. Bend nets are more harmful to the fish, resulting in higher mortality than bag nets do.

Ending the bend net fishery will have a huge impact on fishermen in Finnmark. Dividing the county catch of 2005 on the two types of gear, 29 tons of salmon were caught using 146 bag nets, while 132 tons were caught using 806 bend nets. A change involving the replacement of gear by the more expensive bag nets means investments for those who will still participate in the fishery. The gear replacing bend nets is more work intensive, and it is obvious reasons to believe that some fishermen will not continue their fishing, thereby reducing the importance of the sea-based fishery for salmon.

As the new management signals are crucial to the future sea-based fisheries, arguments for regulations need solid foundation. In principle, the possibility of the sea-based fishery targeting unhealthy stocks is—in the precautionary approach adopted by Norway—sufficient ground to halt this fishery. Regarding the particular rivers, however, specific knowledge on the status of each salmon stock can be developed and used as background for regulatory measures. In accordance with Norway's membership in NASCO Norway adopted the definition of the precautionary approach as implying that only stocks above conservation limits be harvested. The conservation limit was defined as the critical spawning biomass when recruitment shows significant reduction. In order to establish measures enabling managers to state when the critical levels were reached, scientists developed models that estimate the number of eggs per square meters needed to provide recruitment on the level intended (Hindar et al., 2007). The practical approach for managers then, is seen as securing the number of females who optimize the river's potential for production. And managers have qualified a management target to be that the particular spawning stock should be above conservation limits in three out of the four latest years.

This management target represents something new. It gives the salmon a stronger identity in the management regime. The figures on eggs and females are hard-core numbers that appear valid and reliable as proof of a salmon stock condition. But the researchers point out that the data behind the figures to be used in the management scheme for 2008 are highly uncertain. The data stems from stock-recruitment modeling in 9 river systems only, and the researchers label them "first generation spawning biomass targets" (Hindar et al., 2007, p. 3). Expectations are that the data become more solid over the coming years. With solid figures for the salmon stocks in each water system, management measures can be made more accurate and pointed.

The management tool makes promises for a secure future of the salmon stocks. A calculated target figure on sustainability in a certain river will make it easier to regulate the fishery. Failing to fulfill the management targets in a river immediately calls for a halt in that particular fishery. When also reducing the sea-based fishery that potentially harvests stocks below conservation limits, wild Atlantic salmon should have a better future.

User Mobilisation

Although the salmon stocks in Finnmark are generally recognized as healthy, large parts of the sea-based fishery target salmon stocks from the large water system of the Tana River. Several stocks in the Tana River do not meet the management targets for spawning biomass. The fishery also target salmon of Russian origin. The Russians have voiced that the fishery targeting their stocks should stop. Thus, the new regulatory measures will have effects in the Finnmark fishery. And when first suggested, this management change was met with strong opposition from the fishermen, particularly from Finnmark County. A consultation paper was submitted to the interested parties, and opposition to the reference points where voiced by many (Anon., 2008). The Sámi Parliament, The Norwegian fishermen's association, and FeFo—three large, well-organized parties—all criticized the management target for being founded on poor knowledge. They questioned the need to introduce extensive management measures based on weak knowledge. FeFo documented this claim by referring to the fact that local fishermen had applied the scientist's new models on their local river, and reached quite different conclusions as to the number of females that would secure the future of the local salmon stock. FeFo welcomed the new management suggestions as well suited to improve the salmon stocks, but questioned the hastiness of the process introducing them. The Norwegian fishermen's association voiced that there are also other causes than sea-based fisheries that influence salmon stocks, and that focus be directed towards gaining knowledge on such causes. They also argued that the new regulations would be counterproductive towards the target of more wild salmon on the spawning grounds, as reductions in time for sea-based fisheries would reduce catches of escaped farmed salmon that occupied the grounds. They also questioned the knowledge about regulations in general, pointing to the ban on drift nets being introduced in the 1980s, making huge cuts in catches without any sign of improving the stocks of salmon.

In addition to being submitted to a public hearing, the proposed regulations were subject to consultations with the Sámi Parliament. Representing both sea- and river-based fisheries, the Sámi Parliament stated that the fishermen harvesting at sea were now sufficiently regulated. They pointed to the burden carried by the sea-based fishermen for years, and that now was the time for the river-based fishermen to take responsibilities for additional burdens. They retrieved support for this view in the Parliament proposition that opened up for such considerations; "In regions were the effect of reducing catches at sea or in the river have similar effect on the stock in question, the collected consequences for the communities involved must be taken into consideration in allocation of burdens." (St.prp.nr. 32 [2006-2007], pp. 35-36).

Single persons also voiced opinions, and words describing the suggested regulations as "behind all reason," "not thought through properly," and "cultural rape" bear witness not only to colorful language, but of strong emotions relating to the end of the sea-based fishery for Atlantic salmon.

But the regulations also had its supporters. The organizations representing sports fishermen and private owners supported the proposition to a large degree, claiming that the salmon stocks were threatened and strong measures needed. Some river organizations wanted the sea-based fisheries totally banned, and referred to the necessity of securing vulnerable stocks that were part of local sea-based catches. They also referred to neighboring countries having halted their fisheries in the sea and suggested that Norway follow their examples.

Following the consultation period, the new regulations of the salmon fisheries were implemented autumn 2008. They carried rather large restrictions on the sea-based fisheries. The regulations were given each county with additional specifications and details on regulations within zones in that particular county. Ten counties were included in the regulations on salmon fisheries, and four of them had reductions in fishing time with about a month. In five counties the reduction was two weeks, while a single county got reduced fishing time from four to three days a week. It was the best season that was regulated, early summer, around the month of June. As such, the reductions in fishing time for the sea-based fishery are substantial. They are also mainly in accordance with what the Directorate for Nature Management had suggested in the first place – substantial reductions in fishing time during the most attractive fishing period of June. Two exceptions must be mentioned. In Trøndelag and in Finnmark regulations became weaker than the original suggestion made by the Directorate. In some areas in Trøndelag fishermen got to fish 1–2 weeks more than suggested, and with regards to Finnmark, the suggested one month cut in time allowed for fishing with bend nets did not take place. The fishing period was unchanged, but the number of fishing days per week was reduced from four to three.

A possible explanation to the lack of full support for the suggestions of reduced effort is that the interests of sea-based salmon fishermen are better organized in these two counties. The national organization for sea-based salmon fisheries, Norway's Landowner and Sea-Based Salmon Fishermen's Organization, is based in Trøndelag. The organization does not have many members from Finnmark, but the Finnmark fishermen's interests were advocated and spoken for by fishermen's organizations and municipal boards, as well as the Sámi Parliament in consultation with the Directorate for Nature Management. In the original proposal, the Directorate claimed that the interests of the Sámi population were taken sufficiently into consideration. The final result, however, made more concessions to salmon fisheries in Finnmark than originally suggested by the Directorate.

Conclusion

Regulations of the sea-based fishery for salmon in Norway are characterized by the fact that the interested parties do have channels by which they can voice opinions. These channels were also used when regulations for 2008 were to be implemented. The suggested regulations from the management authorities were comprehensive and interested parties voiced opposition. The body for discussions on management issues includes both sea and river interests at one table. Typically, it was the representatives of the sea-based fishermen who voiced opposition to the regulations, while river-based fishermen embraced the suggested reductions on the sea fishery. If the Directorate for Nature Management targets inclusion of all interests and follows user advice in regulatory issues in this situation, its goal is easy to reach. The concerns voiced in the public hearing diverge, and whatever the Directorate for Nature Management decides someone will be pleased. In the case presented here some were more satisfied than others, both at sea and in the rivers. Regulations never became as strong as the private owners wished for, and neither so weak as the sea-based fishermen had hoped for.

The future potential of biologically based management targets as a cornerstone in taking adequate measures in the Norwegian management regime of salmon is interesting. Today's targets are said to be a first generation, but two conditions makes it likely that today's target will be continued and developed. One condition is that managers need simple, practical reference points as foundations for regulations, and are inclined to install regulatory measures even if the full scientific truth about cause and effect is not fully investigated (Smith 1994). The managers' need for scientific bases for regulations also encourages further research. Managers in the Directorate for Nature Management and researchers are teaming up, making a strong alliance in support of biologically based management targets as a cornerstone in the future management of the wild salmon.

A second condition that strengthens the position of scientifically defined management targets relates to their convenient use in the difficult debate between people representing totally different views. The new management targets, which represent the salmon in figures of females and eggs, is fixed in a scientific legitimacy that could silence people arguing that halting the fishery is cultural rape. If so, this leads to questions regarding the future of the Norwegian co-management practices in salmon management: Will the new representations of salmon, salmon seen as eggs per square meter, decrease the relevance of stakeholder involvement? Is it easier to count female salmon than administer people with opposing views?

REFERENCES

Anon. (1992). Lov om laksefisk og innlandsfisk m.v. av 15. mai 1992 nr 47.

Anon. (2007). Bestandsstatus for laks 2007. Rapport fra arbeidsgruppe. Direktoratet for naturforvaltning.

Anon. (2008). Høringsuttalelse – Regulering av fiske etter anadrome laksefisk 2008-2012. Direktoratet for naturforvaltning. Trondheim.

ICES. (1997). *Report of the Working Group on North Atlantic Salmon.* I.C.E.S. C.M. 1997/Assess: 10.

NASCO. (1998). Agreement on Adoption of a Precautionary Approach. *NASCO Doc CNL, (98) 46.*

http://www.nasco.int/pdf/nasco_res_adoptprec.pdf

NOU. (1994). 21. Bruk av land og vann i Finnmark i historisk perspektiv.

NOU. (1999), 9. Til laks åt alle kan ingen gjera?

NOU. (2008). 5. Retten til fiske i havet utenfor Finnmark.

Hindar, K., Diserud, O. H., Fiske, P., Forseth, T., Jensen, A. J., Ugedal, O., Jonsson, N., Sloreid, S. E., Arnekleiv, J. V., Saltveit, S. J., Sægrov, H. & Sættem, L. M. (2007). Gytebestandsmål for laksebestander i Norge. Trondhjem: *NINA Rapport, 226.*

Smith, T. D. (1994). Scaling Fisheries. *The Science of Measuring the Effects of Fishing,* 1855-1955. Cambridge: Cambridge University Press.

St.prp.nr. 32 (2006-2007). Om vern av villaksen og ferdigstilling av nasjonale laksevassdrag og laksefjorder.

In: Salmon: Biology, Nutrition and Consumption
Editors: P. Lacopo et al., pp. 127-143
ISBN: 978-1-60741-287-8

Chapter 4

Dietary Fish Oil Substitution with Rapeseed Oil Alters Tissue Fatty Acid Composition, Liver Lipid Metabolism and Flesh Fat Oxidation in Post-Smolt Atlantic Salmon

D. Menoyo[a*], C.J. Lopez-Bote[b], A. Obach[c] and J.M. Bautista[d]

[a]Departamento de Producción Animal, Universidad Politécnica de Madrid, ETS Ingenieros Agrónomos, 28040, Madrid, Spain

[b]Departamento de Producción Animal, Facultad de Veterinaria, Universidad Complutense de Madrid, 28040 Madrid, Spain

[c]Nutreco Aquaculture Research Centre, N-4001 Stavanger, Norway

[d]Bioquímica y Biologia Molécular IV, Facultad de Veterinaria, Universidad Complutense de Madrid, 28040 Madrid, Spain

Abstract

In the present study, fish oil (FO) was replaced by rapeseed oil (RO) in Atlantic salmon post-smolt diets, and the effects on liver and muscle fatty acid (FA) composition, selected lipogenic and lipolytic enzyme activities and flesh oxidative stability assessed. Fish (220 ± 12g initial BW) were fed five experimental diets for 12 weeks in which the FO was serially replaced by 25, 50, 75, and 100% RO. There were no effects of diet on final weight and biometric indices ($P = 0.118$). The percentage of neutral lipids (NL) linearly decreased in muscle ($P = 0.005$) and increased in liver ($P = 0.001$) with RO inclusion levels. Dietary RO affected the FA composition of muscle and liver NL and polar lipids (PL). In muscle NL, a linear accumulation of total n-6 PUFA was observed with increasing levels of RO ($P < 0.001$), with n-3 HUFA ($P < 0.001$) concentrations decreasing in a linear and quadratic pattern. In liver NL, the concentration of total n-6 PUFA accumulated in a linear and quadratic fashion ($P < 0.001$) while n-3 HUFA linearly decreased ($P < 0.001$) with increasing levels of RO. Total n-3 and n-6 PUFA

* Correspondence: Phone 34-91 452 4854; Fax + 34 91 543 48 79; E-mail david.menoyo@ upm.es

content in the PL fraction were also affected ($P < 0.01$) by dietary input of RO in both liver and muscle. Activity of carnitine palmitoyltransferase I was higher in the liver of fish fed with 100% RO ($P < 0.001$). A time effect ($P < 0.001$) was observed in the concentration of lipid peroxidation products, expressed as thiobarbituric acid reactive substances (TBARS), in fish flesh stored under refrigeration for nine days. Moreover, the progressive inclusion of RO in the feed affected ($P < 0.001$) flesh oxidation stability. In summary, RO can totally replace FO in Atlantic salmon feed producing a decrease in flesh susceptibility to lipid oxidation with no deleterious effect on growth performance. However, a 100% RO diet increased ($P < 0.001$) total n-6 PUFA concentrations and decreased ($P < 0.001$) the concentrations of eicosapentaenoic and docosahexaenoic acids in the fillet.

INTRODUCTION

Salmonids and marine fish culture presently relies heavily on available supplies of fish meal and fish oil (FO) which, because of increased exploitation pressure on strictly limited global feed grade fisheries, as well as the increasing ecological and ethical objections to exploiting a non-sustainable resource, places the industry in a vulnerable position (Sargent and Tacon 1999; Barlow 2000). Therefore, recent research has focused on FO replacement by vegetable oils (VO) in Atlantic salmon diets showing no negative effects on productive parameters (Torstensen et al. 2000; Bell et al. 2001, 2002; Grisdale Helland et al. 2002; Menoyo et al. 2005). Moreover, feeding Atlantic salmon with VO substantially reduces flesh dioxin and dioxin-like PCB concentrations, which is a major concern from a consumer point of view (Bell et al. 2005). According to Bell et al. (2001, 2003a), a suitable VO for fish should contain moderate levels of 18:3n-3 and 18:2n-6, being rich in 18:1n-9. Rapeseed oil (RO) fulfills those characteristics and has been included either blended or alone in several studies performed in Atlantic salmon showing no detrimental effect on growth (Tocher et al. 2000; Bell et al. 2001, 2003a,b, 2005) and gut integrity (Oxley et al. 2005). However, there is a direct relationship between dietary and muscle fatty acid composition in fish (Sargent et al. 1989; Lie et al. 1993) and a replacement of FO with RO at high inclusion level has been shown to decrease 20:5n-3 and 22:6n-3 content in the flesh (Bell et al. 2001; Torstensen et al. 2004a, b). Thus, it is important to study the effects of RO inclusion level on fillet quality by means of the fatty acid composition (neutral and polar lipid fraction) and its susceptibility to lipid peroxidation.

There is increasing concern regarding the possible stimulatory effect of vegetable oils on hepatic elongation and desaturation processes in salmonids (Tocher et al. 2001; Bell et al. 2001; Stubhaug et al. 2005). However, the effect of dietary fatty acid composition on lipogenesis and lipolysis is a controversial matter (Torstensen et al. 2000; Stubhaug et al. 2005, 2007; Menoyo et al. 2005). Therefore, the metabolic use of vegetable oils in salmon feeding still needs to be evaluated. The objectives of this research were (1) to evaluate the influence of dietary fat source (FO vs. RO) and combinations in high-energy diets (36% fat) for Atlantic salmon on muscle and liver fatty acid composition and liver metabolism, and (2) to relate these findings to susceptibility of flesh to lipid peroxidation.

MATERIAL AND METHODS

Fish Husbandry and Feeding

One hundred and ten Atlantic salmon (*Salmo salar*) of average initial weight 220g were randomly distributed into five 500 L polyethylene circular tanks supplied with continuous sea water at the Nutreco Aquaculture Research Center Lerang Research Station, Jørpeland, Norway. The tanks were subjected to a photoperiod regime of 18 h light:6 h dark, and the temperature over the whole experimental period ranged from 8 to 9°C. Fish were fed daily to satiation and feed disappearance was monitored throughout the trial. Wasted feed was collected from the effluent water from each tank by a wire mesh collector and dried. Net feed intake was registered daily. Five practical-type commercial diets were produced at the Nutreco Technology Centre (Stavanger, Norway), as extruded, sinking, 4-mm pellets. Diets were formulated to contain targeted levels of 37% crude protein and 36% crude fat (DM basis). Their basal composition was the same except for the oil added during vacuum fat coating (Table 1). Batches of extruded pellets were produced from a common meal mixture. The pellets were then coated with northern hemisphere fish oil (FO) that was serially replaced by 25, 50, 75, or 100% rapeseed oil (RO). The fatty acid composition of the experimental diets is presented in Table 2.

Biological Parameters and Sample Collection

During the final sampling, 11 fish per tank were killed by a blow to the head and immediately exsanguinated in chilled seawater; weight and fork length were recorded. Then, fish were eviscerated and the weight of viscera and liver registered in order to assess the hepato- and viscero-somatic indexes. Livers were then cut into two pieces, and one portion was placed in liquid N^2 and stored at –80°C for enzymatic analyses, whereas the other half was frozen and stored at –20°C for fatty acid analysis. Fish were filleted and left fillets were immediately vacuum-packaged and frozen at –20°C for fatty acid analysis and lipid peroxidation tests.

Chemical Analysis of Flesh Samples and Lipid Peroxidation Assessment

Fatty acids of diets were extracted and quantified by the one-step procedure described by Sukhija and Palmquist (1988) in freeze-dried samples, with pentadecanoic acid (C15:0; Sigma, Alcobendas, Madrid) as internal standard. Neutral lipids (NL) and polar lipids (PL) from individual fillet and liver samples (five fish per tank) were extracted using the method of Marmer and Maxwell (1981). Before the analysis of fatty acids by gas chromatography, all lipid samples were methylated as described by López-Bote et al. (1997). Fatty acid methyl esters were then analyzed using gas chromatography (Model HP-6890; Hewlet Packard Co., Avondale, PA) equipped with flame ionization detection and a 30 m × 0.32 mm × 0.25 mm cross-linked polyethylene glycol capillary column (HP-Innowax). Results were expressed as the percentage of each fatty acid with respect to the total fatty acids.

Table 1. Composition of experimental diets

	Rapeseed oil, %				
	0	25	50	75	100
Ingredients g/kg DM					
Fish meal-LT (71.8/8.4)	632.0	632.0	632.0	632.0	632.0
Corn gluten	30.1	30.1	30.1	30.1	30.1
Wheat	70.0	70.0	70.0	70.0	70.0
Fish oil	259.4	194.6	129.7	64.9	-
Rapeseed oil	-	64.9	129.7	194.6	259.4
Mineral premix[a]	5.6	5.6	5.6	5.6	5.6
Vitamin premix[a]	2.2	2.2	2.2	2.2	2.2
Carophyll Pink	0.6	0.6	0.6	0.6	0.6
H-tocopherol[b]	333.8	357.4	333.1	336.4	321.4
Htocopherol[b]	12.0	21.7	38.5	51.7	64.8

[a] Active ingredients supplied per kg feed. Minerals: 48 mg $CuSO_4$; 2.0 mg KI; 109.4 mg $MnSO_4$; 0.4 mg Na_2SeO_3; 257.1 mg $ZnSO_4$ (all mineral salts from Sigma, St. Louis). Vitamins: 5mg Retinol acetate; 4.8 mg Vit. D_3; 300 mg H-tocopheryl acetate; 15.3 mg thiamin-Cl; 26 mg riboflavin; 15.3 mg pyridoxine-Cl; 88.9 mg Ca-panthotenate; 153.1 mg niacin; 5.2 mg folic acid; 20.0 mg Vit. B_{12}; 150.0 mg biotin; 408.2 mg myo-inositol; 40.0 mg Vit. K_3; 1g ascorbate polyphosphate (all vitamins from Hoffmann-La Roche).

[b] Analyzed concentration (mg/kg DM)

Table 2. Fatty acid composition of experimental diets (g/100g total fatty acids).

	Rapeseed oil, %				
Fatty acid	0	25	50	75	100
C14:0	6.1	5.1	3.7	2.1	0.8
C16:0	12.0	10.6	9.5	8.3	7.1
C18:0	1.8	1.6	1.8	1.9	2.0
Total SFA	20.1	17.5	15.0	12.5	10.0
C16:1n-7	7.5	6.4	4.6	2.8	1.2
C18:1n-9	13.6	22.5	32.1	42.1	51.0
C20:1n-9	18.5	16.4	12.0	7.3	3.1
C22:1n-11	12.7	11.1	8.0	4.7	1.9
Total MUFA	57.5	61.5	61.4	61.2	61.0
C18:2n-6	3.4	5.8	9.0	12.3	15.3
C20:4n-6	0.4	0.3	0.3	0.2	0.2
Total n-6 fatty acids	4.1	6.4	9.4	12.6	15.6
C18:3n-3	3.6	2.7	4.3	6.1	7.8
C20:5n-3	5.4	4.4	3.5	2.6	1.8
C22:6n-3	6.6	5.3	4.5	0.3	3.0
Total n-3 fatty acids	18.3	14.6	14.1	13.7	13.4
Total n-9 fatty acids	33.9	40.5	45.4	50.3	54.7
n-3:n-6 ratio	4.5	2.3	1.5	1.1	0.9
n-3HUFA[a]	13.0	10.4	8.7	7.1	5.2
UI[b]	1.6	1.5	1.5	1.5	1.5

[a] Highly unsaturated fatty acids (HUFA) includes 20:3, 20:4, 20:5, 22:5 and 22:6.

[b] UI= unsaturation index (average number of double bonds per fatty acid residue).

Lipid peroxidation products were determined as thiobarbituric acid reactive substances (TBARS) in fish flesh according to the procedure of Menoyo et al. (2002), and TBARS were expressed as μmol malonaldehyde (MDA)/kg wet tissue. Vitamin E isoforms were extracted and quantified in feed and muscle as described by Rey et al. (2001).

Mitochondrial Preparations, Soluble Extracts and Enzyme Analyses

Mitochondrial extracts from liver were prepared as described by Menoyo et al. (2004), and the activity of carnitine palmitoyltransferase I (CPT I; EC 2.3.1.21) was assayed according to procedures of Sanz et al. (2000). Liver homogenates and the activity of glucose-

6-phosphate dehydrogenase (G6PD; EC 1.1.1.49) were assayed as described by Bautista et al. (1988). All enzyme activity assays were performed in duplicate or triplicate at 30ºC. Enzymatic activity units (IU), defined as μmoles of substrate converted to product per minute at the assay temperature were expressed per mg of hepatic soluble protein (specific activity). Soluble protein content of liver homogenates was determined by the method of Bradford (1976), using bovine serum albumin (BSA) as the standard. The L – [methyl-^{3}H] carnitine hydrochloride (82.0 Ci/mmol) using in the CPT I determination was supplied by Amersham Pharmacia Biotech (Barcelona, Spain), while Sigma Chemical Co. (Alcobendas, Madrid, Spain) supplied the remaining reagents.

Statistical Analysis

Data were analyzed as completely randomized design by the GLM procedure of SAS (SAS Inst., Inc., Cary, NC), with LO inclusion level as the lone fixed effect, and individual fish as the experimental unit as previously described (Menoyo et al. 2005). The Tukey's test was used to separate treatment means, and regression analysis was used to measure the linear (L) or quadratic (Q) response to RO inclusion level. A repeated measure mean test was used to compare differences in TBARS concentrations between groups during stimulated lipid peroxidation (Morris 1999).

Results and Discussion

There were no significant differences ($P \geq 0.118$) in final weight, length, viscero- and hepato-somatic indexes (Table 3), and also no mortalities were recorded. Similar results were reported by Bell et al. 2001, 2003a in Atlantic salmon fed diets containing up to 25% fat with rapeseed oil as the unique lipid source. Dietary fat type did not affected (P = 0.145) the percentage of polar intramuscular lipids (Table 3). However, the concentration of neutral intramuscular lipids (mainly triglycerides) linearly decreased ($P < 0.005$) with increasing RO inclusion level. This is in agreement with results from Bell et al. (2001), who observed a lower muscle lipid concentration in salmon fed diets containing up to 25% fat with RO replacing more than 50% of FO. Triglycerides are the predominant lipid class in salmon muscle (Bell et al. 2003a) and as the fish increases in size this tissue becomes increasingly important as a fat depot (Jobling and Johansen 2003). Also, it is well reported the increase in muscle adiposity when feeding Atlantic salmon a high-fat (>32% crude fat) diet (Jobling and Johansen 2003). Thus, it is interesting to note that RO decreases the lipid content in the flesh also when included at high levels (36% crude fat), as this not happen with other vegetable oils (Menoyo et al. 2005) and may have important implications in fillet oxidative stability (see below). Intrahepatic NL concentrations linearly increased ($P < 0.001$) with increasing levels of RO in the diet (Table 3). Fatty livers with increasing triglyceride concentrations have been previously described in salmon fed diets containing graded amounts of RO (Bell et al. 2001), linseed oil (Menoyo et al. 2005), soybean oil (Ruyter et al. 2006) or blends of RO and linseed oil (Tocher et al. 2001), but not in fish fed palm oil (Bell et al. 2002) or fed blends of vegetable oils and raised at high (17°C) temperatures (Jordal et al. 2007). Therefore, vegetable oil source and temperature may affect triglyceride accumulation in salmon liver.

Biological availability of dietary fat is directly related to the chemical and physical properties of lipids, including chain length and saturation (Bracco 1994). Salmonids are capable to discriminate among different fatty acids diverting them either toward catabolic (β-oxidation) or synthetic (i.e., fatty acid synthesis and/or elongation-desaturation, phospholipid and triglyceride synthesis) pathways (Kiessling and Keissling 1993; Mckenzie et al. 1998; Bell et al. 2000, 2003b). In a recent study performed in Atlantic salmon hepatocytes, Stubhaug et al. (2005) reported a lower uptake of fatty acids into hepatocytes when feeding a vegetable oil. However, once inside the cell, apparently, instead of being esterified into cellular lipids, they seem to undertake β-oxidation with 18:2n-6 being the most oxidized (Stubhaug et al. 2005). In agreement with these observations, in the present study we have been able to detect a significant (P <0.001) higher specific activity of CPT I (the enzyme who catalyzes the initial step in fatty acid oxidation) in the liver of Atlantic salmon fed 100% RO (Figure 1). The preferential usage of 18:2n-6 as energy substrate in the liver could be partially responsible for the accumulation of 18:1n-9 in the NL fraction (Table 4). This is in accordance with Caballero et al. (2002), who reported lipidic vacuoles as a consequence of 18:1n-9 accumulation in the liver of rainbow trout fed blends of anchovy oil and RO.

Table 3 Effect of dietary fish and rapeseed oil combinations on weight, biometry parameters and muscle and liver neutral (NL) and polar (PL) lipids

	Rapeseed oil (%)					Pooled SE[a]	*P* < F
	0	25	50	75	100		
Weight, g	578.18	597.55	554.36	592.36	522.36	21.44	0.118
Length, cm	33.97	34.02	33.85	34.38	33.38	0.40	0.534
Condition Factor, %	1.47	1.52	1.43	1.46	1.40	0.03	0.176
VSI, %[b]	11.81	11.80	11.67	11.94	11.28	0.31	0.485
HSI, %[c]	1.69	1.60	1.57	1.64	1.51	0.05	0.173
Muscle NL, %[d]	8.60	8.34	6.15	6.61	6.29	0.63	0.005
Muscle PL, %[e]	1.07	0.97	0.98	0.93	1.04	0.04	0.145
Liver NL, %[d]	4.02 [y]	4.65 [y]	4.28 [y]	4.49 [y]	6.10 [z]	0.34	0.001
Liver PL, %[e]	1.42	1.42	1.42	1.37	1.44	0.05	0.926

[a] n = 11 for weight, length, condition factor, VSI and HSI; n = 5 for muscle NL, muscle PL, liver NL and liver PL.

[b] Viscero-somatic index (VSI) = carcass weight x BW $^{-1}$ x 100.

[c] Hepato-somatic index (HSI) = liver weight x BW $^{-1}$ x 100.

[d] NL = neutral lipid.

[e] PL = polar lipid.

[x,y,z] Means within a row that are significantly different are denoted by different superscripts.

No dietary effects were found regarding G6PD activity (Figure 1.). This agrees with Jordal et al. (2007) in Atlantic salmon fed a blend of vegetable oil containing 50% rapessed, but is in contrast to our previous finds on G6PD specific activity in fish fed diets containing variable substitution levels of FO by linseed oil (Menoyo et al. 2005). It has been shown that diets rich in EPA and DHA (FO and/or seaweed), inhibits liver G6PD activity in rats (Murata et al. 2002) but also in rainbow trout hepatocytes (Alvarez et al. 2000). Also the presence of vegetable oils in Atlantic salmon diets enhances elongation processes in liver (Bell et al. 1993; Tocher et al. 2000). Thus, a higher G6PD activity might be expected as vegetable oils replace FO while there is an increasing need of NADPH (supplied by G6PD activity) for the elongation process. The utilization of a Northern hemisphere FO with lower EPA and DHA levels than those of the Southern hemisphere, together with the low levels of 18:3n-3 in the diets, can be behind the different response on G6PD activity found in this study and the one with linseed oil (Menoyo et al. 2005).

Table 4. Effect of dietary fish and rapeseed oil combinations on selected fatty acids (g/100g total fatty acids) of neutral hepatic lipids

	Rapeseed oil (%)					Pooled SE	$P<F$
	0	25	50	75	100		
Fish no.	5	5	5	5	5		
C14:0	2.29^{z}	2.11^{z}	1.59^{y}	1.14^{xy}	0.73^{x}	0.11	<0.001
C16:0	14.20^{z}	10.90^{y}	11.49^{y}	10.17^{xy}	7.68^{x}	0.53	<0.001
C18:0	3.88	3.47	4.01	4.14	3.62	0.20	0.824
Total SFA	20.71^{z}	16.72^{y}	17.28^{y}	15.62^{y}	12.14^{x}	0.64	<0.001
C16:1n-7	3.57^{z}	3.56^{z}	2.58^{y}	1.83^{x}	1.24^{x}	0.19	<0.001
C18:1n-9	16.51^{w}	25.51^{x}	28.30^{xy}	33.99^{y}	43.49^{z}	1.55	<0.001
C20:1n-9	10.02^{yz}	12.09^{z}	8.52^{xy}	6.65^{wx}	6.08^{w}	0.72	<0.001
C22:1n-11	2.80^{w}	2.50^{wx}	1.67^{xy}	1.00^{yz}	0.66^{z}	0.21	<0.001
Total MUFA	37.44^{x}	48.87^{yz}	45.50^{xy}	47.58^{yz}	55.52^{z}	2.56	<0.001
C18:2n-6	2.09^{v}	3.41^{w}	4.83^{x}	6.63^{y}	8.96^{z}	0.17	<0.001
C20:2n-6	0.58^{v}	0.96^{w}	1.21^{x}	1.59^{y}	1.99^{z}	0.02	<0.001
C20:3n-6	0.35^{y}	0.43^{y}	0.90^{z}	1.27^{z}	1.27^{z}	0.10	<0.001
C20:4n-6	1.71^{z}	1.40^{yz}	1.37^{yz}	1.35^{yz}	0.99^{y}	0.11	<0.001
Total n-6 fatty acids	4.73^{v}	6.20^{w}	8.32^{x}	10.97^{y}	13.52^{z}	0.15	<0.001
C18:3n-3	1.53^{y}	0.97^{x}	1.46^{xy}	1.69^{y}	2.43^{z}	0.13	<0.001
C20:5n-3	7.48^{z}	5.09^{y}	4.56^{y}	3.49^{xy}	2.33^{x}	0.43	<0.001
C22:6n-3	23.80^{z}	19.24^{z}	20.13^{z}	18.24^{z}	12.11^{y}	1.51	<0.001
Total n-3 fatty acids	37.12^{z}	28.21^{y}	28.89^{y}	25.83^{xy}	18.82^{x}	1.99	<0.001
Total n-9 fatty acids	27.54^{x}	38.66^{y}	37.76^{y}	41.60^{y}	50.56^{z}	2.15	<0.001
n-3:n-6 ratio	7.86^{z}	4.55^{y}	3.46^{y}	2.35^{x}	1.40^{x}	0.29	<0.001
n-3HUFA[a]	37.26^{z}	28.84^{yz}	29.48^{yz}	26.70^{xy}	18.41^{x}	2.23	<0.001
UI[b]	2.55^{z}	2.21^{xyz}	2.26^{yz}	2.16^{xy}	1.85^{x}	0.09	<0.001

[a]Highly unsaturated fatty acids (HUFA) includes 20:3, 20:4, 20:5, 22:5 and 22:6.

[c]UI= unsaturation index (average number of double bonds per fatty acid residue).

[v,w,x,y,z]Within a row, means lacking a common superscript letter differ ($P<0.05$).

It is generally accepted that dietary fatty acid profile is closely reflected in fatty acid composition of fish tissues (Rosenlund et al. 2001a; Caballero et al. 2002; Bell et al. 2002). Hepatic and intramuscular NL and PL fatty acid compositions are shown in Tables 4, 5, 6 and 7, respectively. There was a direct relationship between the concentration of saturated fatty acids (SAFA) in the diet and the concentration in liver and intramuscular NL fractions.

Table 5. Effect of dietary fish and rapeseed oil combinations on selected fatty acids (g/100g total fatty acids) of polar hepatic lipids

	Rapeseed oil (%)					Pooled SE	$P<F$
	0	25	50	75	100		
Fish no.	5	5	5	5	5		
C14:0	2.09^{z}	2.00^{z}	1.34^{y}	0.99^{x}	0.62^{w}	0.07	<0.001
C16:0	22.59^{z}	20.05^{yz}	20.81^{z}	18.09^{y}	18.01^{y}	0.68	<0.001
C18:0	4.16	3.95	5.23	4.16	5.52	0.63	0.150
Total SFA	29.34^{z}	26.27^{yz}	27.58^{yz}	23.39^{y}	24.29^{y}	1.16	0.001
C16:1n-7	1.85^{z}	1.82^{z}	1.38^{y}	1.15^{y}	0.68^{x}	0.07	<0.001
C18:1n-9	10.49^{w}	13.98^{x}	16.83^{xy}	21.44^{yz}	23.45^{z}	0.02	<0.001
C20:1n-9	5.06^{yz}	6.11^{z}	4.61^{xy}	3.67^{wx}	2.81^{w}	0.30	<0.001
C22:1n-11	0.76^{z}	0.82^{z}	0.50^{y}	0.42^{xy}	0.26^{x}	0.05	<0.001
Total MUFA	20.79^{y}	25.74^{yz}	26.22^{z}	29.52^{z}	29.81^{z}	1.27	<0.001
C18:2n-6	1.63^{v}	2.66^{w}	4.03^{x}	5.83^{y}	7.13^{z}	0.16	<0.001
C20:2n-6	0.52^{w}	0.88^{x}	1.17^{y}	1.57^{z}	1.79^{z}	0.02	<0.001
C20:3n-6	0.45^{x}	0.62^{x}	1.40^{y}	1.92^{yz}	2.53^{z}	0.04	<0.001
C20:4n-6	2.32	2.17	2.21	1.69	2.68	0.34	0.839
Total n-6 fatty acids	4.92^{x}	6.33^{x}	8.81^{y}	11.02^{y}	14.14^{z}	0.02	<0.001
C18:3n-3	0.73^{y}	0.48^{y}	0.71^{y}	1.18^{z}	1.34^{z}	0.08	<0.001
C20:5n-3	11.43^{z}	9.27^{y}	6.70^{y}	5.47^{xy}	4.84^{x}	0.41	<0.001
C22:6n-3	28.86^{z}	28.54^{z}	26.93^{z}	26.45^{z}	23.12^{y}	0.78	<0.001
Total n-3 fatty acids	44.95^{z}	41.66^{z}	37.39^{y}	36.07^{y}	31.76^{x}	0.90	<0.001
Total n-9 fatty acids	15.98^{x}	20.60^{y}	21.99^{y}	25.82^{yz}	26.95^{z}	0.03	<0.001
n-3:n-6 ratio	9.23^{z}	6.58^{y}	4.36^{x}	3.30^{wx}	2.26^{w}	0.31	<0.001
n-3HUFA[a]	46.89^{z}	43.90^{z}	40.20^{y}	38.39^{xy}	35.52^{x}	0.76	<0.001
UI[b]	2.86^{z}	2.78^{z}	2.66^{y}	2.60^{y}	2.44^{x}	0.03	<0.001

[a]Highly unsaturated fatty acids (HUFA) includes 20:3, 20:4, 20:5, 22:5 and 22:6.
[c]UI= unsaturation index (average number of double bonds per fatty acid residue).
[v,w,x,y,z] Within a row, means lacking a common superscript letter differ ($P<0.05$).

However, higher concentrations of SAFA related to diet were found in liver and intramuscular PL showing the relevance of this fatty acid family as membrane structural compounds. Differences in monounsaturated fatty acids (MUFA) concentrations were found in liver despite the similar dietary contents, suggesting a chain length effect of dietary MUFA on liver metabolism (Table 4). Thus, feeding a FO rich in 20:1n-9 and 22:1n-11 promoted less MUFA accumulation in the liver than feeding a RO rich in 18:1n-9 suggesting that although MUFA are readily oxidized in salmon liver, there is a preference for long chain (with more than 20 carbon atoms) MUFA as energy yielding compounds. This effect was also observed in the intramuscular NL fraction (Table 6) as previously reported by Bell et al. 2001, although it was quantitatively less important than in the liver. Increasing levels of RO in the diet increased the MUFA intramuscular contents (mainly as 18:1n-9) in the NL fraction following a linear ($R^2 = 0.72$; $P = 0.003$) and quadratic pattern ($R^2 = 0.72$; $P < 0.001$). Also in the liver NL fraction increased amounts of dietary RO proportion resulted in a linear increase ($R^2 = 0.44$; $P < 0.001$) in the accumulation of MUFA (Table 8). Similar findings regarding MUFA accumulation were observed in the PL fraction in both tissues, but at a lower range because of the quantitative differences of MUFA contents in both lipid fractions (Tables 5 and 7).

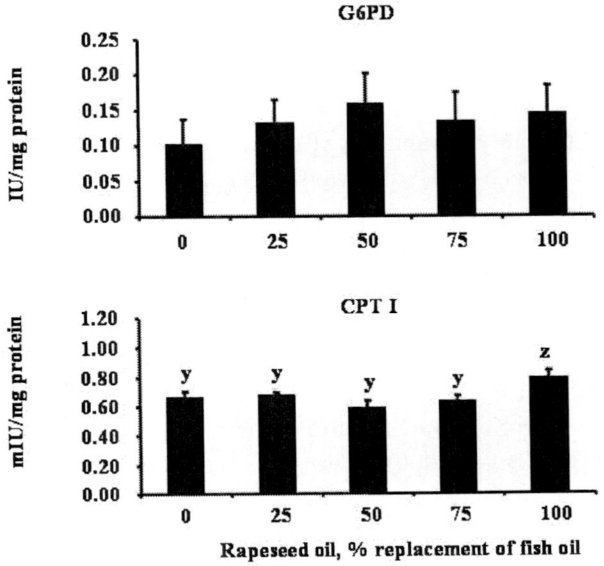

Figure 1. Effect of increasing rapeseed oil in the feed of Atlantic salmon on liver glucose-6-phosphate dehydrogenase (G6PD) and carnitine palmitoyltransferase I (CPT I) activities. Bars lacking a common superscript letter differ ($P < 0.01$).

The concentration of n-6 fatty acids in liver NL and PL were similar and reflected dietary contents (Tables 4 and 5). Thus increasing dietary RO inputs resulted in increasing concentrations of n-6 (mainly as 18:2n-6) in both lipid fractions in a linear ($R^2 = 0.99$; $P < 0.001$) and quadratic ($R^2 = 0.99$; $P < 0.001$) fashion (Table 8). Similar findings were observed in muscle, were a linear increase in NL ($R^2 = 0.90$; $P < 0.001$) and PL ($R^2 = 0.97$; $P < 0.001$) was observed with increasing RO inclusion level. Experimental diets also affected the n-3 HUFA tissue concentration. However, in a similar experiment in which we used a Northern hemisphere FO rich in MUFA but with less n-3 HUFA contents than a Southern hemisphere FO dietary induced effects were quantitatively less important in muscle than in liver (Menoyo et al. 2002). Moreover, while in the i.m NL fraction n-3 HUFA concentration decreased accordingly to dietary contents in a linear ($R^2 = 0.90$; $P < 0.001$) and quadratic ($R^2 = 0.90$; $P < 0.001$) fashion (Table 8), a linear decrease ($R^2 = 0.33$; $P = 0.003$) was observed in the PL fraction richer in longer chain PUFA (Tables 7 and 8). In the liver, FO replacement resulted in a linear decrease in both NL ($R^2 = 0.58$; $P < 0.001$) and PL ($R^2 = 0.85$; $P < 0.001$) fractions.

Increasing levels of RO in the diet promotes a different pattern of individual fatty acid accumulation in liver and muscle PL and NL fractions. According to Bell et al. (2001) a vegetable oil replacing FO should minimize deposition of 18:2n-6, promote 18:3n-3 elongation to maintain relative amounts of EPA and DHA and provide a good source of MUFA as energy yielding compound for growth. In previous studies (Tocher et al. 2000; Bell et al. 2001, 2003b), and in the present one, RO has been proven to fulfill several but not all the conditions to totally replace FO. There seems to be no negative effects on growth, and includes moderate amounts of 18:2n-6 in the fillet. However, because of the low 18:3n-3 contents in RO, the elongation processes are not important enough to maintain high levels of n-3 HUFA in the fillet (Bell et al. 2001) and hence the quality of the fillet is decreased. Moreover, although 18:1n-9 is readily oxidized in the liver, the inclusion of RO with high contents of this fatty acid at high substitution levels promotes its accumulation in the liver and may have a negative long term effect for the fish.

Previous studies performed with Atlantic salmon have shown a lack of an effect of vegetable oils on fillet quality parameters (Rosenlund et al. 2001b; Rørå et al. 2003) and organoleptic attributes (Obach 2001). However, recent studies (Regost et al. 2004) report a firmer raw fillet with lower liquid losses in salmon fed soybean oil or a Peruvian fish oil than fish fed RO or a capelin oil. One of the most important quality attributes in fish is its nutritional value as a source of PUFA (Boggio et al. 1985). However salmon rich in PUFA is more susceptible to lipid oxidation, yielding undesirable lipid oxidation products, which negatively affects quality characteristics. Although it is well known that unsaturated fatty acids of the PL fraction are the main reason for fish flesh peroxidation during storage (Monahan 2000), the susceptibility of flesh to oxidation also depends on the balance of pro- and antioxidants (López-Bote 2000). The effects of experimental diets on TBARS concentration in fish flesh stored under refrigeration for nine days are shown in Table 9. A time effect ($P < 0.001$) on TBARS formation was observed during storage. This is in agreement to literature and confirms the high sensitivity of fish flesh to oxidation along storage (Olsen et al. 1999; Menoyo et al. 2005). A diet effect on TBARS formation was also observed ($P < 0.001$), in which dietary RO inclusion led to a lower oxidation in a quantitative way. This agrees with Regost et al. (2004) who reported lower TBARS values in the raw fillet of salmon fed RO or soybean oil in contrast to fish fed two different sources of fish oil. This is of interest, since lipid oxidation is associated to quality deterioration and accumulation of undesirable products in fish flesh. The lower oxidation in flesh from fish receiving the RO containing diet is probably due to the lower concentration of n-3 HUFA both in NL and PL. Olsen et al. (1999) reported increasing TBARS values in response to increasing dietary PUFA levels in Arctic charr, indications that supplying α-tocopherol above requirements does not improve oxidative stability. Our laboratory previously reported higher TBARS concentration in the flesh of Atlantic salmon fed a diet containing a rich n-3 HUFA FO than when feeding diets formulated with FO that was less rich in n-3 fatty acids (Menoyo et al. 2002). In addition, it is also noticeable that dietary oils provided not only different fatty acid profile but also a marked different supply of γ-tocopherol (Table 1). This led to a significant quantitative response in the concentration of γ-tocopherol in fillets according to the dietary RO inclusion level (Figure 2). This effect was also observed previously in the fillet of salmon fed linseed oil (Menoyo et al. 2005) and soybean and rapeseed oils (Regost et al. 2004). Although α-tocopherol appears to be more active than γ-tocopherol in preventing lipid oxidation along storage, γ-tocopherol has also been proven to have a protective effect (Rey et al. 1998; Ng et al. 2004), thus future research is needed to assess their potential as ante- and postmortem antioxidants in salmon muscle.

Table 6.Effect of dietary fish and rapeseed oil combinations on selected fatty acids (g/100g total fatty acids) of neutral intramuscular lipids

	Rapeseed oil (%)					Pooled SE	P < F
	0	25	50	75	100		
Fish no.	5	5	5	5	5		
C14:0	5.28 ^{z}	4.48 ^{y}	3.44 ^{x}	2.28 ^{w}	1.45 ^{v}	0.05	<0.001
C16:0	13.45 ^{z}	12.03 ^{y}	11.24 ^{x}	10.32 ^{w}	10.00 ^{w}	0.12	<0.001
C18:0	2.63 ^{xy}	2.46 ^{x}	2.57 ^{xy}	2.70 ^{yz}	2.87 ^{z}	0.04	0.001
Total SFA	21.95 ^{z}	19.46 ^{y}	17.66 ^{x}	15.60 ^{w}	14.58 ^{v}	0.16	<0.001
C16:1n-7	6.70 ^{z}	5.70 ^{y}	4.34 ^{x}	2.87 ^{w}	1.93 ^{v}	0.08	<0.001
C18:1n-9	17.74 ^{v}	25.19 ^{w}	32.83 ^{x}	40.61 ^{y}	46.22 ^{z}	0.25	<0.001
C20:1n-9	16.68 ^{z}	15.49 ^{y}	11.65 ^{x}	8.23 ^{w}	5.05 ^{v}	0.26	<0.001
C22:1n-11	9.80 ^{z}	8.68 ^{y}	6.22 ^{x}	3.82 ^{w}	1.91 ^{v}	0.16	<0.001
Total MUFA	56.50 ^{x}	60.70 ^{z}	60.18 ^{yz}	60.20 ^{yz}	58.76 ^{y}	0.42	<0.001
C18:2n-6	3.69 ^{v}	5.44 ^{w}	7.73 ^{x}	9.98 ^{y}	11.76 ^{z}	0.11	<0.001
C20:2n-6	0.48 ^{w}	0.65 ^{x}	0.82 ^{y}	1.12 ^{z}	1.17 ^{z}	0.01	<0.001
C20:3n-6	0.15 ^{w}	0.17 ^{w}	0.23 ^{x}	0.31 ^{y}	0.41 ^{z}	0.02	<0.001
C20:4n-6	0.30 ^{z}	0.26 ^{y}	0.23 ^{xy}	0.20 ^{x}	0.22 ^{x}	0.01	<0.001
Total n-6 fatty acids	4.62 ^{v}	6.51 ^{w}	9.02 ^{x}	11.62 ^{y}	13.57 ^{z}	0.13	<0.001
C18:3n-3	3.34 ^{x}	2.09 ^{w}	3.12 ^{x}	4.01 ^{y}	4.57 ^{z}	0.19	<0.001
C20:5n-3	3.26 ^{z}	2.48 ^{y}	2.07 ^{x}	1.67 ^{w}	1.52 ^{w}	0.08	<0.001
C22:6n-3	6.68 ^{z}	5.87 ^{y}	5.27 ^{xy}	4.52 ^{w}	4.55 ^{wx}	0.17	<0.001
Total n-3 fatty acids	16.93 ^{z}	13.33 ^{y}	13.14 ^{y}	12.58 ^{y}	13.10 ^{y}	0.32	<0.001
Total n-9 fatty acids	36.14 ^{v}	42.31 ^{w}	45.84 ^{x}	49.94 ^{y}	51.61 ^{z}	0.34	<0.001
n-3:n-6 ratio	3.67 ^{z}	2.05 ^{y}	1.46 ^{x}	1.08 ^{w}	0.97 ^{w}	0.06	<0.001
n-3HUFA[a]	12.93 ^{z}	10.84 ^{y}	9.70 ^{y}	8.47 ^{x}	8.41 ^{x}	0.26	<0.001
UI[b]	1.48 ^{z}	1.41 ^{y}	1.41 ^{y}	1.41 ^{y}	1.45 ^{yz}	0.01	<0.001

[a]Highly unsaturated fatty acids (HUFA) includes 20:3, 20:4, 20:5, 22:5 and 22:6.
[c]UI= unsaturation index (average number of double bonds per fatty acid residue).
^{v,w,x,y,z} Within a row, means lacking a common superscript letter differ (P < 0.05).

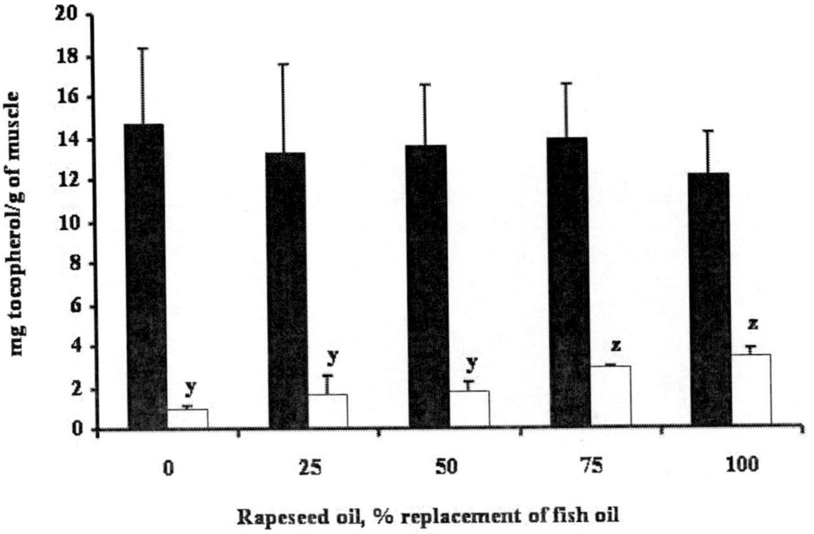

Figure 2. Effect of dietary rapeseed oil inclusion level 0, 25, 50, 75, and 100% on muscle tissue (mg/g muscle) vitamin E isomers concentration. Bars lacking a common superscript letter differ (P < 0.01).

Table 7. Effect of dietary fish and rapeseed oil combinations on selected fatty acids (g/100g total fatty acids) of polar intramuscular lipids

	Rapeseed oil (%)					Pooled SE	$P<F$
	0	25	50	75	100		
Fish no.	5	5	5	5	5		
C14:0	3.25^{z}	3.20^{z}	2.31^{y}	1.59^{xy}	0.96^{w}	0.13	<0.001
C16:0	16.75^{z}	16.85^{z}	16.10^{yz}	14.55^{xy}	14.17^{x}	0.39	<0.001
C18:0	3.14^{yz}	1.70^{x}	2.00^{x}	2.40^{xy}	3.71^{z}	0.25	<0.001
Total SFA	23.56^{z}	22.16^{yz}	20.75^{xy}	18.78^{x}	19.04^{x}	0.50	<0.001
C16:1n-7	3.65^{z}	3.37^{z}	2.52^{y}	1.67^{x}	1.08^{x}	0.15	<0.001
C18:1n-9	11.36^{v}	15.57^{w}	19.74^{x}	23.87^{y}	27.62^{z}	0.70	<0.001
C20:1n-9	8.06^{z}	7.32^{z}	5.61^{y}	3.97^{x}	2.58^{w}	0.02	<0.001
C22:1n-11	3.57^{z}	3.23^{yz}	2.38^{y}	1.46^{x}	0.81^{x}	0.21	<0.001
Total MUFA	30.48	33.41	33.97	34.49	35.50	1.43	0.041
C18:2n-6	2.33^{w}	3.84^{x}	5.30^{y}	7.21^{z}	8.17^{z}	0.02	<0.001
C20:2n-6	0.38^{w}	0.53^{x}	0.70^{y}	1.01^{z}	1.07^{z}	0.02	<0.001
C20:3n-6	0.22^{w}	0.26^{w}	0.40^{x}	0.62^{y}	0.88^{z}	0.02	<0.001
C20:4n-6	0.93^{z}	0.67^{y}	0.63^{y}	0.64^{y}	0.75^{y}	0.03	<0.001
Total n-6 fatty acids	3.86^{w}	5.30^{x}	7.02^{y}	9.48^{z}	10.95^{z}	0.02	<0.001
C18:3n-3	2.37^{y}	1.79^{x}	2.51^{y}	3.32^{z}	3.65^{z}	0.17	<0.001
C20:5n-3	8.16^{z}	7.47^{yz}	6.35^{xy}	5.50^{wx}	4.56^{w}	0.29	<0.001
C22:6n-3	27.07	26.16	25.80	24.02	23.04	1.52	0.063
Total n-3 fatty acids	42.11	39.14	38.26	37.26	34.52	1.64	0.007
Total n-9 fatty acids	20.24^{x}	23.69^{xy}	26.06^{yz}	28.45^{z}	30.74^{z}	1.04	<0.001
n-3:n-6 ratio	10.93^{z}	7.39^{y}	5.50^{x}	3.95^{wx}	3.15^{w}	0.37	<0.001
n-3HUFA[a]	40.27	38.03	36.28	34.04	32.09	1.81	0.003
UI[b]	2.71	2.62	2.60	2.55	2.46	0.09	0.092

[a]Highly unsaturated fatty acids (HUFA) includes 20:3, 20:4, 20:5, 22:5 and 22:6.
[c]UI= unsaturation index (average number of double bonds per fatty acid residue).
[v,w,x,y,z] Within a row, means lacking a common superscript letter differ ($P<0.05$).

Table 8. Effects of different levels of dietary rapeseed oil (RO, % of total added fat) on fatty acid classes (g/100g total fatty acids) in muscle and liver neutral (NL) and polar (PL) lipids

Fatty acid		Intercept	RO	RO^2	R^2	Pooled SE	P-value Linear	P-value Quadratic
Muscle NL								
Saturated	y =	21.95	-0.104	+0.0003	0.98	0.16	<0.001	<0.001
Monounsaturated	y =	56.92	+0.139	-0.0012	0.72	0.42	0.003	<0.001
n-6	y =	4.46	+0.092		0.99	0.13	<0.001	-
n-3HUFA[a]	y =	12.92	-0.091	+0.0004	0.9	0.26	<0.001	<0.001
Muscle PL								
Saturated	y =	23.33	-0.05		0.71	0.50	<0.001	-
Monounsaturated	y =	31.45	+0.04		0.18	1.43	0.040	-
n-6	y =	3.62	+0.07		0.97	0.19	<0.001	-
n-3HUFA	y =	40.2	-0.08		0.33	1.81	0.003	-
Liver NL								
Saturated	y =	20.14	-0.072		0.77	0.64	<0.001	-
Monounsaturated	y =	40	+0.139		0.44	2.56	<0.001	-
n-6	y =	4.66	+0.058	+0.0003	0.99	0.15	<0.001	<0.001
n-3HUFA	y =	36.1	-0.159		0.58	2.23	<0.001	-
Liver PL								
Saturated	y =	28.76	-0.05		0.35	1.16	0.001	-
Monounsaturated	y =	22.05	+0.08		0.56	1.27	<0.001	-
n-6	y =	4.86	+0.05	+0.0003	0.94	0.36	<0.001	0.03
n-3HUFA	y =	46.63	-0.11		0.85	0.85	<0.001	-

[a]Highly unsaturated fatty acids (HUFA) includes 20:3, 20:4, 20:5, 22:5 and 22:6

Table 9. Effect of dietary rapeseed oil inclusion level on thiobarbituric reactive substances (TBARS; µmol of malonyl dialdehyde [MDA] /kg of wet tissue), concentration along fish flesh stored under refrigeration for 9 d.

	Rapeseed oil, %					Pooled SE[1]	Pooled SE[2]	Probability of contrasts		
	0	25	50	75	100			RO[3]	time	time x RO
Storage time (d)										
0	0.11	0.12	0.18	0.09	0.09	0.14	0.04	<0.001	<0.001	<0.001
3	0.54	0.43	0.53	0.23	0.37					
6	1.12	0.88	1.03	0.49	0.47					
9	1.83	1.03	1.29	0.66	0.73					

[1]Pooled SE of the dietary treatments (n=5).
[2]Pooled SE of the mean of time and interaction time x treatments (n=20).
[3]RO, rapeseed oil %

Conclusion

Dietary rapeseed oil can totally replace fish oil in Atlantic salmon feed, producing a decrease in flesh susceptibility to lipid oxidation with no deleterious effect on growth performance. However, fatty acid composition in muscle and liver is transformed, leading to an increase in n-6 polyunsaturated fatty acids and a reduction of n-3 highly unsaturated fatty acid concentration in fish flesh when fed diets higher in rapeseed oil.

Acknowledgments

This research was partially financed by the Ministerio de Ciencia y Tecnología (CICYT AGL-2004-06958). The authors are grateful to R. Prieto for technical assistance.

References

Alvarez, M. J., Díez, A., López-Bote, C., Gallego, M. & Bautista, J. M. (2000). Short-term modulation of lipogenesis by macronutrients in rainbow trout (*Oncorhynchus mykiss*) hepatocytes. *British Journal of Nutrition, 84*, 619-628.

Barlow, S. (2000). Fishmeal and fish oil: sustainable ingredients for aquafeeds. *Global Aquaculture Advocate, 4*, 85-88.

Bautista, J. M., Garrido-Pertierra, A. & Soler, G. (1988). Glucose-6-phosphate dehydrogenase from *Dicentrarchus labrax* liver: kinetic mechanism and kinetic of NADPH inhibition. *Biochimica et Biophysica Acta, 967*, 354-363.

Bell, J. G., Dick, J. R. & Sargent, J. R. (1993). Effects of diets rich in linoleic or α-linolenic acid on phospholipid fatty acid composition and eicosanoid production in Atlantic salmon (*Salmo salar*). *Lipids, 28*, 819-826.

Bell, J. G., McEvoy, J., Tocher, D. R. & Sargent, J. R. (2000). Depletion of α-tocopherol and astaxanthin in Atlantic salmon (*Salmo salar*) affects autoxidative defense and fatty acid metabolism. *Journal of Nutrition, 130*, 1800-1808.

Bell, J. G., McEvoy, J., Tocher, D. R., McGhee, F., Campbell, P. J. & Sargent, J. R. (2001). Replacement of fish oil with rapeseed oil in diets of Atlantic salmon (Salmo *salar*) affects tissue lipid composition and hepatocyte fatty acid metabolism. *Journal of Nutrition, 131*, 1535-1543.

Bell, J. G., Henderson, R. J., Tocher, D. R., McGhee, F., Dick, J. R., Porter, A., Smullen, R. P. & Sargent, J. R. (2002). Substituting fish oil with crude palm oil in the diet of Atlantic salmon (*Salmo salar*) affects muscle fatty acid composition and hepatic fatty acid metabolism. *Journal of Nutrition*, *132*, 222-230.

Bell, J. G., McGhee, F., Campbell, P. J. & Sargent, J. R. (2003a). Rapeseed oil as an alternative to marine fish oil in diets of post-smolt Atlantic salmon (*Salmo salar*): changes in flesh fatty acid composition and effectiveness of subsequent fish oil "wash out". *Aquaculture*, *218*, 515-528.

Bell, J. G., Tocher, D. R., Henderson, R. J., Dick, J. R. & Crampton, V. O. (2003b). Altered fatty acid compositions in Atlantic salmon (*Salmo salar*) fed diets containing linseed and rapeseed oils can be partially restored by a subsequent fish oil finishing diet. *Journal of Nutrition*, *133*, 2793-2801.

Bell, J. G., McGhee, F., Dick, J. R. & Tocher, D. R. (2005). Dioxin and dioxin-like polychlorinated biphenyls (PCBs) in Scottish farmed salmon (*Salmo salar*): effects of replacement of dietary fish oil by vegetable oils. *Aquaculture*, *243*, 305-314.

Bracco, U. (1994). Effect of triglyceride structure on fat absorption. *American Journal of Clinical Nutrition*, *60*, 1002S-1009S.

Boggio, S. M., Hardy, R. W., Babbitt, J. K. & Brannon, E. L. (1985). The influence of dietary lipid source and alpha-tocopheryl acetate level on product quality of rainbow trout (*Salmo gairdneri). Aquaculture*, *51*, 13-24.

Bradford, M. M. (1976). A rapid and sensitive method for the quantitation of microgram quantities of protein utilizing the principle of protein-dye binding. *Analytical Biochemistry*, *72*, 248-254.

Caballero, M. J., Obach, A., Rosenlund, G., Montero, D., Gisvold, M. & Izquierdo, M. S. (2002). Impact of different dietary sources on growth, lipid digestibility, tissue fatty acid composition and histology of rainbow trout, *Oncorhynchus mykiss*. *Aquaculture*, *214*, 253-271.

Grisdale-Helland, B., Ruyter, B., Rosenlund, G., Obach, A., Helland, S. J., Gisvold, M., Standal, H. & Røsjø, C. (2002). Influence of high contents of dietary soybean oil on growth, feed utilization, tissue fatty acid composition, heart histology and standard oxygen consumption of Atlantic salmon raised at two temperatures. *Aquaculture*, *207*, 311-329.

Jobling, M. & Johansen, S. J. S. (2003). Fat distribution in Atlantic salmon *Salmo salar* L. in relation to body size and feeding regime. *Aquaculture Research*, *34*, 311-316.

Jordal, A. E. O, Lie, Ø. & Torstensen, B. (2007). Complete replacement of dietary fish oil with a vegetable oil blend affect liver lipid and plasma lipoprotein levels in Atlantic salmon (Salmo salar L.). *Aquaculture Nutrition*, *13*, 114-130.

Kiessling, K. H. & Keissling, A. (1993). Selective utilization of fatty acids in rainbow trout (*Oncorhynchus mykiss* Walbaum) red muscle mitochondria. *Canadian Journal of Zoology*, *71*, 248-251.

Lie, Ø., Sandvin, A. & Waagbø, R. (1993). Influence of dietary fatty acids on the lipid composition of lipoproteins in farmed Atlantic salmon (Salmo salar). *Fish Physiology and Biochemistry*, *12*, 249-260.

López-Bote, C. J., Rey, A. I., Sanz, M., Gray, J. I. & Buckley, D. J. (1997). Dietary vegetable oils and α-tocopherol reduce lipid oxidation in rabbit muscle. *Journal of Nutrition, 127*, 1176-1182.

López-Bote, C. J. (2000). Dietary treatment and quality characteristics in mediterranean meat products. In: E., Decker, E. A., Faustman, & C. J. Lopez-Bote, (Eds.), *Antioxidants in Muscle Foods* (1st edition, pp. 345-366). New York, NY: Wiley & Sons, Inc.

Marmer, W. N. & Maxwell, R. J. (1981). Dry column method for the quantitative extraction and simultaneous class separation of lipids from muscle tissue. *Lipids*, *16*, 365-371.

Mckenzie, D. J., Higgs, D. A., Dosanjh, B., Deacon, G. & Randall, D. J. (1998). Dietary lipid composition influences swimming performance in Atlantic salmon (*Salmo salar*) in seawater. *Fish Physiology and Biochemistry*, *19*, 111-122.

Murata, M., Sano, Y., Ishihara, K. & Ushida, M. (2002). Dietary fish oil and Undaria pinnatifida (wakame) synergistically decrease rat serum and liver triacylglycerol. *Journal of Nutrition*, *132*, 742-747.

Menoyo, D., López-Bote, C. J., Bautista, J. M. & Obach, A. (2002). Herring vs. anchovy fish oils in salmon feeding. *Aquatic Living Resources*, *15*, 217-223.

Menoyo, D., López-Bote, C. J., Bautista, J. M. & Obach, A. (2003). Growth, digestibility and fatty acid utilization in large Atlantic salmon (*Salmo salar*) fed varying levels of n-3 and saturated fatty acids. *Aquaculture*, *225*, 295-307.

Menoyo, D., Izquierdo, M. S., Robaina, L., Ginés, R., Lopez-Bote, C. J. & Bautista, J. M. (2004). Adaptation of lipid metabolism, tissue composition and flesh quality in gilthead sea bream (Sparus aurata) to the replacement of dietary fish oil by linseed and soyabean oils. *British Journal of Nutrition*, *92*, 41-52.

Menoyo, D., López-Bote, C. J., Obach, A. & Bautista, J. M. (2005). Effect of dietary fish oil substitution with linseed oil on the performance, tissue fatty acid profile, metabolism, and oxidative stability of Atlantic salmon. *Journal of Animal Science*, *83*, 2853-2862.

Monahan, F. J. (2000). Oxidation of lipids in muscle foods: fundamentals and applied concerns. In: E., Decker, E. A., Faustman & C. J. Lopez-Bote, (Eds.), *Antioxidants in Muscle Foods* (1st edition, pp. 3-24). New York, NY: Wiley & Sons, Inc.

Morris, T. R. (1999). *Experimental design and analysis in animal sciences*. CAB International OX10 8DE UK.

Obach, A., Bendiksen, E. Å., Rosenlund, G. & Gisvold, M. (2001). Impact of dietary lipid source on muscle fatty acid composition and sensory evaluation of Atlantic salmon (*Salmo salar* L.). In: S. C. Kestin, & P. D. Warriss, (Eds), *Farmed Fish Quality*, (1st edition, pp. 391-392) Fishing News Books, *Blackwell Science Ltd.*

Olsen, R. E., Løvaas, E. & Lie, Ø. (1999). The influence of temperature, dietary polyunsaturated fatty acids, α-tocopherol and spermine on fatty acid composition and indices of oxidative stress in juvenile Arctic char, (*Salvelinus alpinus* L.). *Fish Physiology and Biochemistry*, *20*, 13-29.

Oxley, A., Torstensen, B. E., Rustan, A. C. & Olsen, R. E. (2005). Enzyme activities of intestinal triacylglycerol and phosphatidylcholine biosynthesis in Atlantic salmon (*Salmo salar* L.). *Comparative Biochemistry and Physiology B*, *141*, 77-87.

Regost, C., Jakobsen, J. V. & R ørå, A. M. B. (2004). Flesh quality of raw and smoked fillets of Atlantic salmon as influenced by dietary oil sources and frozen storage. *Food Research International*, *37*, 259-271.

Rey, A. I., Isabel, B., Cava, R. & Lopez-Bote, C. J. (1998). Dietary acorns provide a source of gamma-tocopherol to pigs raised extensively. *Canadian Journal of Animal Science*, *78*, 441-443.

Rey, A. I., Kerry, J. P., Lynch, P. B., Lopez-Bote, C. J. & Buckley, Morrissey, D. J. (2001). Effect of dietary oils and α-tocopheryl acetate supplementation on lipid (TBARS) and cholesterol oxidation in cooked pork. *Journal of Animal Science*, *79*, 1201-1208.

Rosenlund, G., Obach, A., Bendiksen, E. Å., Gisvold, M. & Ruyter, B. (2001a). Effect of dietary fatty acid composition on Atlantic salmon (*Salmo salar*) when replacing fish oils with vegetable oils. In: S.C. Kestin & P.D. Warriss (Eds), *Farmed Fish Quality*, (1st edition, 403-404) Fishing News Books, Blackwell Science Ltd.

Rosenlund, G., Obach, A., Sandberg, M. G., Standal, H. & Tveit, K. (2001b). Effect of alternative lipid sources on long-term growth performance and quality of Atlantic salmon (*Salmo salar*). *Aquaculture Research*, *32*(Suppl. 1), 1-7.

Rørå, A. M. B., Regost, C. & Lampe, J. (2003). Liquid holding capacity, texture and fatty acid profile of smoked fillets of Atlantic salmon fed diets containing fish oil or soybean oil. *Food Research International*, *36*, 231-239.

Ruyter, B., Moya-Falcon, C., Rosenlund, G. & Vegusdal, A. (2006). Fat content and morphology of liver and intestine of Atlantic salmon (Salmo salar): Effects of temperature and dietary soybeanoil. *Aquaculture*, *252,* 441-452.

Sanz, M., Lopez-Bote, C., Menoyo, D. & Bautista, J. M. (2000). Abdominal fat deposition and fatty acid synthesis are lower and β-oxidation is higher in broiler chickens fed diets containing unsaturated rather than saturated fat. *Journal of Nutrition*, *130*, 3034-3037.

Sargent, J. R., Henderson, R. J. & Tocher, D. R., (1989). The lipids. In: J. E. Halver (Ed), *Fish Nutrition*, (1st Edition, 154-218). Academic Press, New York.

Sargent, J. R. & Tacon, A. G. J. (1999). Development of farmed fish: a nutritionally necessary alternative to meat. *Proceedings of the Nutrition Society*, *58*, 377-383.

Stubhaug, I., Tocher, D. R., Bell, J. G., Dick, J. R. & Torstensen, B. E. (2005). Fatty acid metabolism in Atlantic salmon (Salmo salar L.) hepatocytes and influence of dietary vegetable oil. *Biochimica et Biophysica Acta*, *1734*, 277-288.

Stubhaug, I., Lie, Ø. & Torstensen, B. E. (2007). Fatty acid productive value and β-oxidation capacity in Atlantic salmon (Salmo salar L.) fed on different lipid sources along the whole growth period. *Aquaculture Nutrition*, *13*, 145-155.

Sukhija, P. S. & Palmquist, D. L. (1988). Rapid method for determination of total fatty acid content and composition of feedstuffs and feces. *Journal of Agriculural and Food Chemistry*, *36*, 1202-1206.

Tocher, D. R., Bell, J. G., Dick, J. R., Henderson, R. J., McGee, F., Michell, D. & Morris, P. C. (2000). Polyunsaturated fatty acid metabolism in Atlantic salmon (*Salmo salar*) undergoing parr-smolt transformation and the effects of dietary linseed and rapeseed oils. *Fish Physiology and Biochemistry*, *23*, 59-73.

Tocher, D. R., Bell, J. G., MacGlaughlin, P., McGee, F. & Dick, J. R. (2001). Hepatocyte fatty acid desaturation and polyunsaturated fatty acid composition of liver in salmonids: effects of dietary vegetable oil. *Comparative Biochemistry and Biophysics*, *B130*, 257-270.

Torstensen, B. E., Lie, Ø. & Frøyland, L. (2000). Lipid metabolism and tissue composition in Atlantic salmon (*Salmo salar* L.).- effects of capelin oil, palm oil, and oleic acid-enriched sunflower oil as dietary lipid sources. *Lipids*, *35*, 653-664.

Torstensen, B. E., Frøyland L. & Lie. Ø. (2004a). Tailoring of a cardioprotective muscle fatty acid composition of Atlantic salmon (Salmo salar) fed vegetable oils. *Food Chemistry*, *87*, 567-580.

Torstensen, B. E., Frøyland L. & Lie. Ø. (2004b). Replacing dietary fish oil with increasing levels of rapeseed oil and olive oil—effects on Atlantic salmon (*Salmo salar* L.) tissue and lipoprotein lipid composition and lipogenic enzyme. *Aquaculture Nutrition*, *10*, 175-192.

In: Salmon: Biology, Nutrition and Consumption
Editors: P. Lacopo et al., pp. 145-157
ISBN: 978-1-60741-287-8

Chapter 5

POLYCYCLIC AROMATIC HYDROCARBON OCCURRENCE IN FRESH SALMONIDS

Vincent Varlet **and Thierry Serot***
Laboratoire de Biochimie Alimentaire et Industrielle LBAI ENITIAA – UMR CNRS GEPEA, 6144 Rue de la Géraudière, BP 82225
44322 Nantes, Cédex 3, France. Fax: 0033 251 785 520

ABSTRACT

Polycyclic aromatic hydrocarbons (PAHs) constitute carcinogenic environmental contaminants such as dioxins. They are found in all biotopes of living organisms. Fish, particularly salmonids, are very sensitive to PAH toxicity. The natural and anthropogenic generation of PAHs lead to environmental contamination of water, sediments and feed of salmonids, then to fish contamination and finally to human food contamination when the affected fish are eaten. Therefore, they are responsible for an initial food contamination because they are present in raw food material. The monitoring of these chemical contaminants in salmon is essential to evaluate pollution related to human activities and to guarantee the quality of fish as food for human consumption.

I. POLYCYCLIC AROMATIC HYDROCARBONS

I.1. PAHs as Environmental Contaminants

Polycyclic aromatic hydrocarbons (PAHs) constitute a class of chemical compounds made up of two to six (sometimes more) benzene fused rings. They are produced by incomplete burning of organic material due to natural or/and human activity. As result, PAHs

* Corresponding author: E-mail vincentvarlet@laposte.net; Phone 0033 611 506 531

constitute environmental contaminants. The monitoring of several of them has been recommended since the 1970s by the United States Environmental Protection Agency (US EPA), which established a list of 16 environmental PAHs, compiled in Table 1 (Veyrand et al., 2007).

Later, PAHs were classified in the Persistent Organic Pollutants (POPs) category of chemical contaminants as dioxins, furans and several pesticides thanks to the Aarhus protocol in 1998, which came into force in 2003 (Wenzl et al., 2006). This directive was revised in 2001 by Decision 2001/752/EC, and benzo[a]pyrene (B[a]P), the most famous and the most studied PAH, was proposed as the leading substance in PAH contamination (EC, 2001). As POPs, PAHs are identified by four main properties: bioaccumulation, toxicity for human health, environmental persistence and capacity to travel significant distances.

Table 1. Polycyclic aromatic hydrocarbons whose monitoring was recommended by the US EPA in the 1970s (US EPA, 1999)

PAH	Structure	Molecular weight
Naphthalene		128 a.m.u.
Acenaphthylene	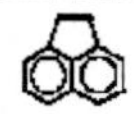	152 a.m.u.
Acenaphthene	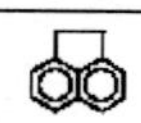	154 a.m.u.
Fluorene	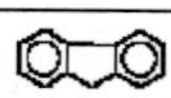	166 a.m.u.
Phenanthrene	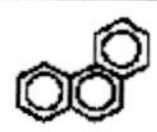	178 a.m.u.
Anthracene	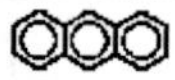	178 a.m.u.
Fluoranthene	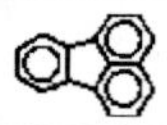	202 a.m.u.
Pyrene	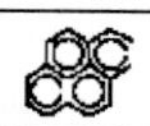	202 a.m.u.
Benzo[a]anthracene	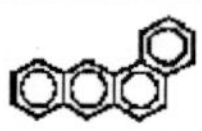	228 a.m.u.
Chrysene	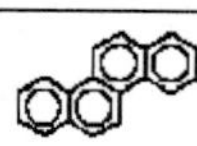	228 a.m.u.
Benzo[b]fluoranthene	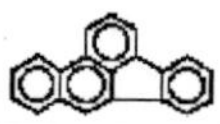	252 a.m.u.
Benzo[k]fluoranthene	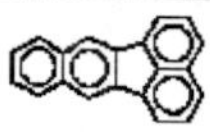	252 a.m.u.
Benzo[a]pyrene	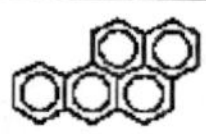	252 a.m.u.
Indeno[1,2,3-c,d]pyrene	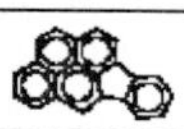	276 a.m.u.
Benzo[g,h,i]perylene		276 a.m.u.
Dibenzo[a,h]anthracene	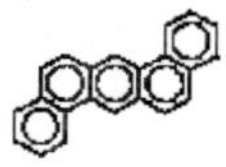	278 a.m.u.

Table 2. Polycyclic aromatic hydrocarbons whose monitoring was recommended by the European Union (EC, 2005a)

PAH	Structure	Molecular weight
Benzo[c]fluorene	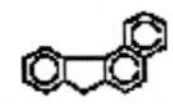	216 a.m.u.
Cyclopenta[c,d]pyrene	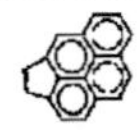	226 a.m.u.
Benzo[a]anthracene	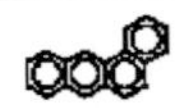	228 a.m.u.
Chrysene	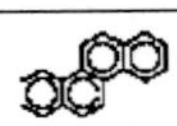	228 a.m.u.
5-methyl chrysene	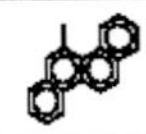	246 a.m.u.
Benzo[b]fluoranthene	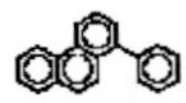	252 a.m.u.
Benzo[j]fluoranthene	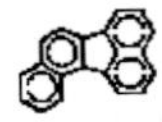	252 a.m.u.
Benzo[k]fluoranthene	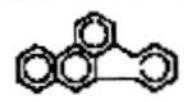	252 a.m.u.
Benzo[a]pyrene		252 a.m.u.
Indeno[1,2,3-c,d]pyrene	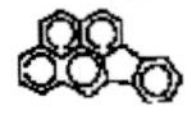	276 a.m.u.
Benzo[g,h,i]perylene		276 a.m.u.
Dibenzo[a,h]anthracene	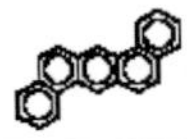	278 a.m.u.
Dibenzo[a,e]pyrene	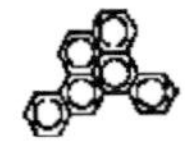	302 a.m.u.
Dibenzo[a,h]pyrene	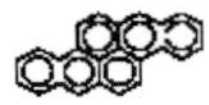	302 a.m.u.
Dibenzo[a,i]pyrene	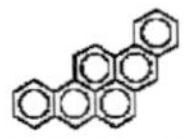	302 a.m.u.
Dibenzo[a,l]pyrene		302 a.m.u.

PAHs constitute very apolar substances (log K_{ow} range between 4 and 7) and, thanks to their hydrophobicity, they can cross lipidic biological membranes and accumulate in tissues. Therefore, an accumulation of PAHs in the food chain is noticeable and their concentrations increase until the end of the chain.

PAHs are classified as carcinogenic compounds and are shown responsible for causing defects on human health (Baird et al., 2005). They are metabolized to dihydrodiols by hepatic hydrocarbon hydroxylases. A part of these metabolites is excreted via urine, bile and fecal matter, but another part of these dihydrodiols and respective epoxides derivatives can bind to DNA and proteins, causing mutagenic processes (Varlet, 2007a).

Due to their molecular stability, the degradation of PAHs is not easy. They resist strong alkaline or acidic treatments and, as dioxins, they have a half-life of several years in atmosphere, especially PAHs with heavy molecular weight. For lighter PAHs, UV can accelerate their degradation as certain bacterial enzymes.

Finally, mainly thanks to their cyclic structure and hydrophobicity, PAHs can be adsorbed easily in particulates and contaminate places very far from those of human activity.

Therefore, PAHs are responsible for an initial raw food material contamination. Since the 1990s, several European and American scientific opinions have advocated monitoring these environmental contaminants (EC, 2002).

I.2. PAHs as Food Contaminants

PAHs can also be considered food contaminants. Indeed, additional food industrial processes can lead to the formation of PAHs and increase the initial contamination. The use of processes such as smoking or flavouring with wood smoke condensates is responsible for PAH food contamination because PAHs, generated during wood pyrolysis, are found sometimes in important quantities in the final product. Therefore, the European Union adopted new legislation in 2005 that provided a list of 15 PAHs (8 in common with the US EPA environmental PAH contaminants list), presented in Table 2. The monitoring of these PAHs was recommended due to their toxic properties and consequences on human health (EC, 2005a). This recommendation was following by a regulation 208/2005/EC setting new maximum limits for B[a]P in several food categories (oils, baby food, smoked meat, muscle meats other than smoked, crustaceans, bivalve molluscs) (EC, 2005b).

Thus, in such processed food, PAH contamination is often attributed to the industrial process. However, the total PAH contamination is the sum of the environmental and industrial process contamination. The measurement of environmental PAH occurrence in processed food matrices is often forgotten, which is the reason it is essential to monitor the initial PAH concentrations in fresh food matrices.

II. Initial PAH Contamination of Salmonids

II.1. Use of Salmonids as Matrix for the Monitoring of PAH Toxicity

Fish, particularly salmonids, are very sensitive to PAH toxicity, particularly at their early life development stages (Bailey et al., 1996; Woźny et al., 2008). Recent studies suggest that pollution from PAHs can affect growth and reproductive viability in salmon populations (Spromberg et al., 2005) and salmonids in general. Many studies that tend to demonstrate in vivo and in vitro PAH toxicity in salmonids are available.

For in vitro evaluation of PAH toxicity, cultured cells of salmonids, especially rainbow trout, are commonly used (Lannan et al., 1984): the photocytotoxicity of fluoranthene metabolites was assessed on gill epithelial cells (Schirmer et al., 1997), B[a]P genotoxicity was investigated in normal primary hepatocytes (Tsuji et al., 2007), the direct cytotoxicity of 16 PAHs was shown in a cell line from the rainbow trout gill (Schirmer et al., 1998), etc.

For in vivo assessment of PAH toxicity, to maximise the occurrence of PAHs in fish organisms, direct intraperitoneally injections are carried out rather than diet supplementations or water contamination. According to the monitored toxic effects, different kinds of salmonids are used.

- Juvenile immature rainbow trout (mainly *Oncorhyncus mykiss*) are employed to study the influence of PAH on cytochrome mRNA expression in liver (Woźny et al., 2008), on liver metabolism and plasma cortisol levels (Tintos et al., 2007 and 2008), on brain metabolism and monoaminergic neurotransmitters (Gesto et al., 2006 and 2008; Brzuzan et al., 2007).
- Adult salmonids are used to investigate hormonal troubles consequently to PAH injections. The effects of PAHs on energy metabolism in liver and gonads and plasma steroid hormones were investigated in rainbow trout (Tintos et al., 2006). Adult female coho salmon were used to study the PAH influence on sex steroid secretion in plasma (Afonso et al., 1997).

Therefore, salmonids constitute very reactive organisms towards PAH toxicity. This sensitivity explains the use of salmonids as biomarkers and the necessity to investigate PAHs in raw fish intended for human consumption.

II.2. Use of Salmonids as PAH Pollution Biomarkers

Taking into account the bioaccumulation property of PAHs in salmonids, several studies have investigated PAH environmental contamination using salmonids as biomarkers of PAH occurrence (De Flora et al., 1993). A compilation of PAH bioaccumulation in fish (organs, tissues, etc.) including salmonids has been recently carried out (Kirby et al., 2000 ; Van Der Oost et al., 2003).

PAHs and/or derivatives metabolites were searched in the bile and liver of trout (*Oncorhyncus mykiss*) or Chinook salmon (*Oncorhyncus tshawyttscha*) that had lived or had been caged in polluted aquatic media such as the Biobío River in Chile (Barra et al., 2001; Inzunza et al., 2006) or Puget Sound estuaries in the United States (Stein et al., 1995).

To monitor the PAH contamination of the fish, two approaches are possible: direct measurements of PAH concentrations in organs and tissues, and measurement of some biological parameters in contaminated organisms by comparison with noncontaminated fishes.

The direct measurement of PAHs in organs or tissues can provide fast information about the aquatic contamination, but the results are not very reproducible. Indeed, the quantity of PAHs in accumulating organs such liver are very random and specific to each fish even among a species. Liver is the first organ to accumulate PAHs due to its detoxification activity. It has been used as a PAH biomarker, but recent studies have shown a large variability of PAH contents among individual salmon of a same batch (Barbee et al., 2008; Payne et al., 2003). As result, the liver seems not to be a good biomarker of PAH exposure or effect. The analysis of PAHs in the stomach is also very specific to each fish and its dietary habits. However, it can provide information about the direct PAH contamination of the fish diet.

The measurements of biological parameters such as cytochrome P4501A activity must be preferred to direct measurement of PAH concentrations. This cytochrome activity has been investigated in juvenile Chinook salmon of Fraser River in Canada (Wilson et al., 2000) or in trout of the Biobío River in Chile (Barra et al., 2001), both rivers suspected of PAH pollution.

Other biological parameters can also be monitored. The level of DNA binding in hepatic and/or gill tissues can also illustrate the influence of PAHs on the genetic material of salmonids (Stein et al., 1995; Barbee et al., 2008). Micronuclei formation in erythrocytes and chromosome damages in peripheral blood constitute also indicators of PAH toxicity in juvenile coho salmon (*Oncorhyncus kisutch*) (Barbee et al., 2008).

The monitoring of these biological parameters appears as more reliable to PAH pollution than direct PAH measurements.

II.3. PAH Occurrence in Salmon Considered as Human Food Raw Material

PAHs can easily be absorbed and/or ingested by salmonids. However, salmonids can be reared or fished for human consumption. Therefore, PAH monitoring in salmonids, considered a human food raw material, is essential to guarantee the safety of the food.

Thus, other investigations have been undertaken to measure the PAH concentrations in salmons intended for human consumption before food industrial processing. This measurement allows illustrating the environmental PAH contamination as well as distinguishing PAHs produced by environmental contamination from PAHs generated during industrial processes.

The PAH contamination found in wild or reared fresh salmon bought in the supermarket or obtained from farms is presented in Table 3. The occurrence of environmental PAHs (PAHs with a low molecular weight) is generally higher than "food PAHs" (PAHs with a number of benzene fused rings up to four). However, the main critical point is the important variability among the raw salmon. Indeed, salmon reared in Norway (Varlet et al., 2007b), in Vancouver and the Sidney regions (Easton et al., 2001), salmon obtained from a supermarket in Spain (Aguinaga et al., 2008) and salmon from Norway and the Irish Sea (Pena et al., 2006; Visciano et al., 2006) show very different PAH concentrations. Sometimes, the B[a]P concentration in raw salmon is higher than the half of the maximum limit of 5 µg/kg allowed in the same smoked fish (EC, 2006; Visciano et al., 2006). Another study has shown the concentrations of 27 individual PAHs in the raw muscles of trout from Canadian Great Lakes. They ranged from below the detection limit (0.05 µg/kg) to 1.24 µg/kg in lean fish and to 3.86 µg/kg in fat trout (Zabik et al., 1996; Stołyhwo et al., 2005).

Thus, it is necessary to understand and monitor the different origins of PAHs in salmon in order to decrease their occurrence as much as possible in fresh fish.

Table 3. Polycyclic aromatic hydrocarbons recovered in several kinds of salmon

PAH	concentration ± standard deviation (ng/g of dry matter)	Type of fish	References
Naphthalene	0.58 ± 0.08	Fresh salmon	Aguinaga et al., 2008
	2.15	Wild salmon	Easton et al., 2002
	1.98	Farmed salmon	
Acenapthylene	0.15	Wild salmon	Easton et al., 2002
	0.35	Farmed salmon	
Acenaphthene	1.10	Wild salmon	Easton et al., 2002
	0.83	Farmed salmon	
	12.39 ± 7.26	Raw salmon	Visciano et al., 2006
Fluorene	0.61	Wild salmon	Easton et al., 2002
	0.36	Farmed salmon	
	0.06 ± 0.041 [a]	Farmed salmon	Varlet et al., 2007
	18.38 ± 5.75	Raw salmon	Visciano et al., 2006
Phenanthrene	3.1 ± 0.01	Fresh salmon	Aguinaga et al., 2008
	2.34	Wild salmon	Easton et al., 2002
	0.81	Farmed salmon	
	0.04 ± 0.018 [a]	Farmed salmon	Varlet et al., 2007
	26.44 ± 8.09	Raw salmon	Visciano et al., 2006
Anthracene	0.15	Wild salmon	Easton et al., 2002
	0.16	Farmed salmon	
	0.01 ± 0.005 [a]	Farmed salmon	Varlet et al., 2007
	2.81 ± 0.95	Raw salmon	Visciano et al., 2006
Fluoranthene	0.69 ± 0.001	Fresh salmon	Aguinaga et al., 2008
	1.16	Wild salmon	Easton et al., 2002
	0.4	Farmed salmon	
	0.075 ± 0.100 [a]	Farmed salmon	Varlet et al., 2007
	57.45 ± 7.28	Raw salmon	Visciano et al., 2006
Pyrene	0.37	Wild salmon	Easton et al., 2002
	0.11	Farmed salmon	
	0.19 ± 0.015 [a]	Farmed salmon	Varlet et al., 2007
Benzo[a]anthracene	0.01	Farmed salmon	Easton et al., 2002
	< 0.02	Farmed salmon	Varlet et al., 2007
	97.28 ± 20.38	Raw salmon	Visciano et al., 2006
Cyclopenta[c,d]pyrene	< 0.01 [a]	Farmed salmon	Varlet et al., 2007
Chrysene	0.06	Wild salmon	Easton et al., 2002
	0.12	Farmed salmon	
	0.01 ± 0.03 [a]	Farmed salmon	Varlet et al., 2007
	8.74 ± 8.22	Raw salmon	Visciano et al., 2006
Benzo[b]fluoranthene	< 0.02 [a]	Farmed salmon	Varlet et al., 2007
	3.03 ± 3.53	Raw salmon	Visciano et al., 2006
Benzo[j]fluoranthene	< 0.02 [a]	Farmed salmon	Varlet et al., 2007
Benzo[k]fluoranthene	< 0.02 a	Farmed salmon	Varlet et al., 2007
	0.64 ± 0.73	Raw salmon	Visciano et al., 2006
Benzo[a]pyrene	< 0.02 [a]	Farmed salmon	Varlet et al., 2007
	3.67 ± 3.99	Raw salmon	Visciano et al., 2006
Benzo[g,h,i]perylene	< 0.01 [a]	Farmed salmon	Varlet et al., 2007
	0.93 ± 2.52	Raw salmon	Visciano et al., 2006

[a]: Concentration in ng/g of wet tissue

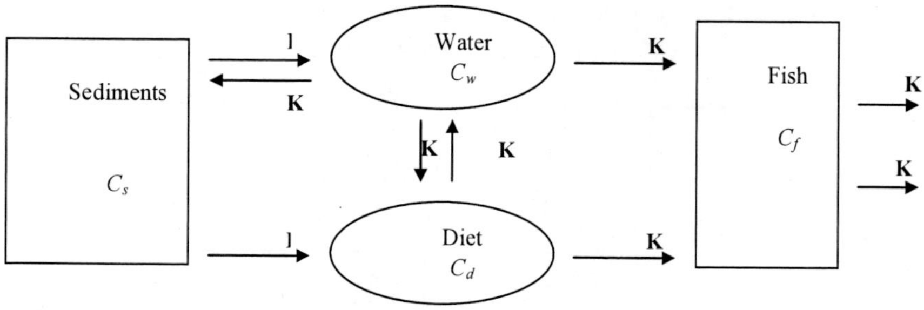

C refers to a concentration and **K** to a rate constant.
S, d, w, ex and met refer, respectively to sediments, diet, water, excretion and metabolic activity.

Figure 1. Simplified bioaccumulation model of PAH in aquatic organisms (adapted from Van Der Oost et al., 2003).

III. Origins of Raw Fish PAH Contamination

The important variability of PAH contamination of salmon is directly related to the environmental contamination (Lemke et al., 1997; Van Der Oost et al., 2003). PAHs may accumulate in salmonids through different mechanisms: via the direct uptake from water by gills or skin (bioaccumulation), via uptake of suspended particles (ingestion) and via the consumption of contaminated food (biomagnification). PAHs of low molecular weight are the most prevalent PAHs in the environment. It is normal to find them as predominant in raw salmonids (Visciano et al., 2008). Due to their less important size, they are more easily bioaccumulated or ingested than PAHs with heavier molecular weight. As salmonids have a high fat matter rate, the bioaccumulation of lipophilic PAHs is not negligeable and more important than in fish with a weaker lipid content. A bioaccumulation model of PAHs in aquatic organisms is shown in Figure 1. PAHs generated from natural and/or human activites contaminate the biotope (sediments, water) and diet of salmonids. The implified model presents the difficulty to consider individually these three contamination routes because they constitute media in constant interactions. Moreover, others parameters not shown in the model can influence the relations between sediments, water, feed and fishes such as salinity, feed ration level, water temperature and age of the fish (Seubert et al., 1997 and 2000; Kennedy et al., 2004). Finally, another kind of PAH contamination comes from the contaminated fish itself. Indeed, by its metabolism and excretion activities, contaminants can be rejected in the aquatic environment (Figure. 1). The contamination of an inland Alaska lake by the migration of contaminated sockeye salmon *Oncorhyncus nerka* has been already demonstrated (Ewald et al., 1998).

III.1. PAH Contamination in Sediments and Water

They are two types of marine sediment PAH contamination: anthropogenic and natural processes. Both sources generate petrogenic PAHs and pyrogenic PAHs.

Marine sediments are naturally contaminated by PAH through biogeochemical cycles. These natural petrogenic PAHs derive from diagenesis of organic material and sediments. The use of crude and refined petroleum obtained from these naturally contaminated sediments lead to anthropogenic petrogenic PAHs. They are introduced in aquatic media through accident oil spills, discharge from routine tanker operations, municipal and urban runoff, (Liu et al., 2008).

The initial PAH contamination of marine sediments can be also attributed for a weaker part to natural pyrogenic PAHs. These PAHs are generated through natural combustion of organic material during wood fires and volcanic activity. Finally, the last route of PAH contamination of sediments is the combustion of fossil fuels (petroleum, coal, etc.) and biomass which produce anthropogenic pyrogenic PAHs. These PAHs can directly contaminate the water or be adsorbed on particulates before sedimentation. Therefore, the PAH contamination of sediments is a consequence of the PAH contamination of seawater. However, seawater contamination (and latter sediment contamination) derives also from freshwater contamination. As human industrial activities are related to the freshwater cycle, chemical contaminants are accumulated in rivers and lakes and finally in estuaries (Thomas et al., 2002). According to the streams, the freshwater contamination is disseminated in oceans and accumulated in marine sediments.

It must be noted that rivers carry also important natural nutrients which make estuaries key ecosystems in which numerous fish species including salmonids live. Then, fishing activities and farms are often located near contaminated areas (Wade et al., 2008) and could explain a part of the initial PAH contamination of salmonids.

Thus, PAH sediment contamination is strongly subjected to human activities (Kose et al., 2008). By way of urbanisation, industrial emissions and environmental accidents, an important variability of PAH occurrence is put into evidence. Because of the physico-chemical equilibria between the water in which salmonids live and the contaminated sediments, PAHs are ultimately accumulated by fish. Water appears to be the key pool of PAHs, which can be stored in sediments or directly in the fish through bioaccumulation or ingestion.

III.2. Feed Contamination

The PAH contamination of water and sediments leads to contamination of salmonids. However, salmonids live in the same medium as their diet. Environmental contamination is also responsible for the occurrence of PAHs in salmonid feed and, consequently, in salmonids. However, except in accidental cases, PAH contamination of wild salmonids through PAH-contaminated feed is low, especially if salmonids are fished far from the coasts.

The PAH contamination of salmon by its diet must be especially considered for farmed salmon. In British Columbia, benthic organisms around salmon pens were found to contain unusually high levels of contaminants. Further investigation demonstrated that the source of this contamination was the salmon feed (Easton et al., 2001). Indeed, the fish oil and fish meal used by the manufacturers of salmon feed could come from various and numerous marine food chains in the world. They are susceptible contamination by variable levels of PAHs (Liguori et al., 2006). As PAHs are lipophilic, the use of uncertified fish oil and fish meal in salmon feed increases the concentrations of such contaminants in the human food chain.

Conclusion

Due to their physico-chemical properties, PAHs contaminate the environment and diet of salmonids and, by extent, human food raw material. As salmonids show a high fat matter rate, lipophilic PAHs can accumulate easily, which explains the use of salmonids as biomarkers of environmental PAH contamination and as study material to understand PAH toxicity. The mutagenic properties of DNA and protein adducts of PAH metabolites have proven the necessity to monitor PAH contamination in fresh salmonids to avoid an increase of fish mortality as well as contamination of human food. Special monitoring of PAHs in the diet and the aquatic environment must be carried out to prevent and control PAH contamination of salmonids.

References

Afonso, L. O. B., Campbell, P. M., Iwama, G. K., Devlin, R. H. & Donaldson, E. M. (1997). The effect of the aromatase inhibitor fadrozole and two polynuclear aromatic hydrocarbons on sex steroid secretion by ovarian follicles of coho salmon. Gen. Comp. *Endocrinol.*, *106*, 169-174.

Aguinaga, N., Campillo, N., Viñas, P. & Hernández-Córdoba, M. (2008). Evaluation of solid-phase microextraction conditions for the determination of polycyclic aromatic hydrocarbons in aquatic species using gas chromatography. *Anal. Bioanal. Chem.*, *391*, 1419-1424.

Bailey, G. S., Williams, D. E. & Hendricks, J. D. (1996). Fish models for environmental carcinogenesis: the rainbow trout. Environ. *Health Perspect.*, *104*, 5-21.

Baird, W. M., Hooven, L. A. & Mahadevan, B. (2001). Carcinogenic polycyclic aromatic hydrocarbon-DNA adducts and mechanism of action. Environ. Mol. *Mutat.*, *45*, 106-114.

Barbee, G. C., Barich, J., Duncan, B., Bickham, J. W., Matson, C. W., Hintze, C. J., Autenrieth, R. L., Zhou, G. D., McDonald, T. J., Cizmas, L., Norton, D. & Donnelly, K. C. (2008). *In situ* biomonitoring of PAH-contaminated sediments using juvenile coho salmon (*Oncorhyncus kisutch*). Ecotox. Environ. *Safety*, in press.

Barra, R., Sanchez-Hernandez, J. C., Orrego, R., Parra, O. & Gavilan, J. F. (2001). Bioavailability of PAHs in the Biobio River (Chile): MFO activity and biliary fluorescence in juvenile *Oncorhyncus mykiss*. *Chemosphere*, *45*, 439-444.

Brzuzan, P., Woźny, M. & Łuczyński, M. Góra. (2007). CYP1A gene expression in adipose fin of rainbow trout (*Oncorhyncus mykiss* Walbaum) exposed to benzo[a]pyrene. *Environ. Biotechnol.*, *3*, 20-24.

De Flora, S., Vigano, L., D'Agostini, F., Camoirano, A., Bagnasco, M., Bennicelli, C., Melodia, F. & Arillo, A. (1993). Multiple genotoxicity biomarkers in fish exposed in situ to polluted river water. Mutat. *Res.*, *319*, 167-177.

Easton, M. D. L., Luszniak, D. & Von Der Geest, E. (2001). Preliminary examination of contaminant loadings in farmed salmon, wild salmon and commercial salmon feed. *Chemosphere*, *46*, 1053-1074.

EC. 2001/752. (2001). Commission Decision of 17 October 2001 amending the Annexes to Council Decision 97/101/EC establishing a reciprocal exchange of information and data from networks and individual stations measuring ambient air pollution within the Member States. *Official Joiurnal of European Union, L 282*, 69-76.

EC. (2002). Opinion of the Scientific Committee on Food on the Risks to Human Health of Polycyclic Aromatic Hydrocarbons in Food. (http://ec.europa.eu/food/fs/sc/scf/out153_en.pdf)

EC. 2005/108. (2005a). Commission Recommendation of 4 February 2005 on the further investigation into the levels of polycyclic aromatic hydrocarbons in certain foods. *Official Journal of the European Union, L 34*, 43-45.

EC. 208/2005. (2005b). Commission Regulation (EC) No 208/2005 of 4 February 2005 amending Regulation (EC) No 466/2001 as regards polycyclic aromatic hydrocarbons. *Official Journal of the European Union, L, 34*, 3-5.

EC. 1881/2006. (2006). Commission Regulation of 19 December 2006 setting maximum levels for certain contaminants in foodstuffs. *Official Journal of the European Union, L 364*, 5-24.

Ewald, G., Larsson, P., Linge, H., Okla, L. & Szarzi, N. (1998). Biotransport of organic pollutants to an inland Alaska lake by migrating sockeye salmon (*Oncorhyncus nerka*). *Arctic*, *51*, 40-47.

Gesto, M., Tintos, A., Soengas, J. L. & Míguez, J. M. (2006). Effects of acute and chronic naphthalene exposure on brain monoaminergic neurotransmitters in rainbow trout (*Oncorhyncus mykiss*). Comp. Biochem. *Physiol. C*, *144*, 173-183.

Gesto, M., Soengas, J. L. & Míguez, J. M. (2008). Acute and prolonged stress responses of brain monoaminergic activity and plasma cortisol levels in rainbow trout are modified by PAHs (naphthalene, beta-naphthoflavone and benzo(a)pyrene) treatment. Aquat. *Toxicol.*, *86(3)*, 341-351.

Inzunza, B., Orrego, R., Peñalosa, M., Gavilán, J. F. & Barra, R. (2006). Analysis of CYP4501A1, PAHs metabolites in bile, and genotoxic damage in *Oncorhyncus mykiss* exposed to Biobío River sediments, Central Chile. Ecotoxicol. *Environ. Safety*, *65*, 242-251.

Kennedy, C. J., Higgs, D. & Tierney, K. (2004). Influence of diet and ration level on benzo[a]pyrene metabolism and excretion in rainbow trout (*Oncorhyncus mykiss*). Arch. Environ. Contam., *Toxicol.*, *47*, 379-386.

Kirby, M. F., Lyons, B. P., Waldock, M. J., Woodhead, R. J., Goodsir, F., Law, R. J., Matthiessen, P., Neall, P., Stewart, C., Thain, J. T., Tylor, T. & Feist, S. W. (2000). Biomarkers of polycyclic aromatic hydrocarbon (PAH) exposure in fish and their application in marine monitoring. Science Series Technical Report Number 110, Center for Environment, *Fisheries and Aquaculture Science (CEFAS)*, Lowestoft.

Kose, T., Yamamoto, T., Anegawa, A., Mohri, S. & Ono, Y. (2008). Source analysis for polycyclic aromatic hydrocarbons in road dust and urban runoff using marker compounds. *Desalination*, *226*, 151-159.

Lannan, C. N., Winton, J. R. & Fryer, J. L. (1984). Fish cell lines: establishment and characterization of nine cell lines from salmonids. *In Vitro*, *20*, 671-676.

Lemke, M. A. & Kennedy, C. J. (1997). The uptake, distribution and metabolism of benzo[a]pyrene in coho salmon (*Oncorhyncus kisutch*) during the parr-smolt transformation. Environ. *Toxicol. Chem.*, *16*, 1384-1388.

Liguori, L., Heggstad, K., Hove, H. T. & Julshamn, K. (2006). An automated extraction approach for isolation of 24 polyaromatic hydrocarbons (PAHs) from various marine matrixes. *Anal. Chim. Acta*, 573-574, 181-188.

Liu, Y., Chen, L., Jianfu, Z., Qinghui, H., Zhiliang, Z. & Hongwen, G. (2008). Distribution and sources of polycyclic aromatic hydrocarbons in surface sediments of rivers and an estuary in Shanghai, China. *Environ. Poll.*, *154(2)*, 298-305.

Payne, J. F., Mathieu, A. & Collier, T. K. (2003). Ecotoxicological studies focusing on marine and freshwater fish. In: *PAHs: An Ecotoxicological Persepctive*, P. E. T. Douben, (Ed.), Wiley, Hoboken, NJ, United States, 191-224.

Pena, T., Pensado, L., Casais, C., Mejuto, C., Phen-Tan-Luu, R. & Cela, R. (2006). Optimization of a microwave-assisted extraction method for the analysis of polycyclic aromatic hydrocarbons from fish samples. *J. Chromatogr. A*, *1121*, 163-169.

Schirmer, K., Chan, A. G. J., Greenberg, B. M., Dixon, D. G. & Bols, N. C. (1997). Methodology for demonstrating and measuring the photocytotoxicity of fluoranthene to fish cells in culture. Toxicol. *In Vitro*, *11*, 107-119.

Schirmer, K., Dixon, D. G., Greenberg, A. G. J. & Bols, N. C. (1998). Ability of 16 priority PAHs to be directly cytotoxic to a cell line from the rainbow trout gill. *Toxicol.*, *127*, 129-141.

Seubert, J. M. & Kennedy, C. J. (1997). The toxicokinetics of benzo[a]pyrene in juvenile coho salmon *Oncorhyncus kisutch*, during smoltification. Fish Physiol. *Biochem.*, *16*, 437-447.

Seubert, J. M. & Kennedy, C. J. (2000). Benzo[a]pyrene toxicokinetics in rainbow trout (*Oncorhyncus mykiss*) acclimated to different salinities. Arch. Environ. Contam. *Toxicol.*, *38*, 342-349.

Spromberg, J. A. & Meador, J. P. (2005). Relating results of chronic toxicity responses to population-level effects: modelling effects on wild Chinook salmon populations. Integr. Environ. *Assess. Manage.*, *1*, 9-21.

Stein, J. E., Hom, T., Collier, T. K., Brown, D. W. & Varanasi, U. (1995). Contaminant exposure and biochemical effects in outmigrant juvenile Chinook salmon from urban and nonurban estuaries of Puget Sound, Washington. Environ. Toxicol. *Chem.*, *14*, 1019-1029.

Stołyhwo, A. & Sikorski, Z. E. (2005). Polycyclic aromatic hydrocarbons in smoked fish – a critical review. *Food Chem.*, *91*, 303-311.

Thomas, K. V., Balaam, J., Barnard, N., Dyer, R., Jones, C., Lavender, J. & McHugh, M. (2002). Characterisation of potentially genotoxic compounds in sediments collected from United Kingdom estuaries. *Chemosphere*, *49(3)*, 247-258.

Tintos, A., Gesto, M., Míguez, J. M. & Soengas, J. L. (2006). Interactive effects of naphthalene treatment and the onset of vitellogenesis on energy metabolism in liver and gonad, and plasma steroid hormones of rainbow trout *Oncorhyncus mykiss*. Comp. *Biochem. Physiol. C*, *144*, 155-165.

Tintos, A., Gesto, M., Míguez, J. M. & Soengas, J. L. (2007). Naphthalene treatment alters liver intermediary metabolism and levels of steroid hormones in plasma of rainbow trout (*Oncorhyncus mykiss*). *Ecotoxicol. Environ. Safety*, *66*, 139-147.

Tintos, A., Gesto, M., Míguez, J. M. & Soengas, J. L. (2008). B-naphthoflavone and benzo[a]pyrene treatment affect liver intermediary metabolism and plasma cortisol levels in rainbow trout *Oncorhyncus mykiss*. Ecotoxicol. *Environ. Safety*, *69*, 180-186.

Tsuji, P. A. & Walle, T. (2007). Benzo[a]pyrene-induced cytochrome P4501A and DNA binding in cultured trout hepatocytes – Inhibition by plant polyphenols. *Chemico-Biolobical Interactions*, *169*, 25-31.

United States Environmental Protection Agency, (1999). *Compendium method* TO-13A, EPA, Cincinatti, OH, USA.

Van Der Oost, R., Beyer, J. & Vermeulen, N. P. E. (2003). Fish bioaccumulation and biomarkers in environmental risk assessment: a review. *Environ. Toxicol. Pharmacol.*, *13*, 57-149.

Varlet, V. (2007a). Caractérisation des composés volatils responsables des qualités odorantes du saumon fumé (Salmo salar) et évaluation des contaminants du fumage (Hydrocarbures Aromatiques Polycycliques). *Thèse de doctorat*, Université de Nantes, France (In French).

Varlet, V., Serot, T., Monteau, F., Le Bizec, B. & Prost, C. (2007b). Determination of PAH profiles by GC-MS/MS in salmon processed by four cold-smoking techniques. *Food Addit. Contamin.*, *24(7)*, 744-757.

Veyrand, B., Brosseaud, A., Sarcher, L., Varlet, V., Monteau, F., Marchand, P., Andre, F. & Le Bizec, B. (2007). Innovative method for determination of 19 polycyclic aromatic hydrocarbons in food and oil samples using gas chromatography coupled to tandem mass spectrometry based on an isotope dilution approach. *J. Chromatogr. A*, *1149*, 333-344.

Visciano, P., Perugini, M., Amorena, M. & Ianeri, A. (2006). Polycyclic aromatic hydrocarbons in fresh and cold-smoked Atlantic salmon fillets. *J. Food Prot.*, *69(5)*, 1134-1138.

Visciano, P., Perugini, M., Conte, F. & Amorena, M. (2008). Polycyclic aromatic hydrocarbons in farmed rainbow trout (*Oncorhyncus mykiss*) processed by traditional flue gas smoking and by liquid smoke flavourings. *Foood Chem. Toxicol.*, *46*, 1409-1413.

Wade, T. L., Sweet, S. T. & Klein, A. G. (2008). Assessment of sediment contamination in Casco Bay, Maine, *USA. Environ. Poll.*, *152*, 505-521.

Wenzl, T., Simon, R., Kleiner, J. & Anklam, E. (2006). Analytical methods for polycyclic aromatic hydrocarbons (PAHs) in food and the environment needed for new food legislation in the European Union. Trends Anal. *Chem.*, *25(7)*, 716-725.

Wilson, J. Y., Addison, R. F., Martens, D., Gordon, R. & Glickman, B. (2000). Cytochrome P4501A and related measurements in juvenile Chinook salmon (*Oncorhyncus tshawytscha*) from the Fraser River. Can. *J. Fish. Aquat. Sci.*, *57*, 405-413.

Woźny, M., Brzuzan, P., Łuczyński, M., Góra, M., Bidzińska, J. & Jurkiewicz, P. (2008). Effects of cyclopenta[c]phenanthrene and its derivatives on *zona radiata* protein, ERα, and CYP1A mRNA expression in liver of rainbow trout (*Oncorhyncus mykiss* Walbaum). *Chemico-Biological Interactions*, *174*, 60-68.

Zabik, M. E., Booren, A., Zabik, M. J., Welch, R. & Humphrey, H. (1996). Pesticides residues, PCBs and PAHs in baked, charbroiled, salt boiled and smoked Great Lakes lake trout. *Food Chem.*, *55(3)*, 231-239.

In: Salmon: Biology, Nutrition and Consumption
Editors: P. Lacopo et al., pp. 159-168

ISBN: 978-1-60741-287-8

Chapter 6

HIGH PRESSURE PROCESSING OF SALMON

Hainer Wackerbarth
German Institute of Food Technology, Prof.-von-Klitzing-Str. 7,
D-49610 Quakenbrück

ABSTRACT

Important changes in food consumption and distribution systems during the last 50 years have been related to the industrialization of food processing. Salmon, initially produced in salt-dried, cold-smoked form by simple tradesmen, was originally regarded as a luxury product, which has now become an industrial product (Cardinal, 2001). While the introduction of aquaculture of the Atlantic salmon has increased the consumption of smoked salmon, problems related to quality still persist. Even there does not exist standardized criteria for the shelf-life, the commercial shelf-life is usually limited to only one week when chill-stored.

Non-thermal food preservation in particular high-pressure processing becomes more applied in the industries. Prolonging of the shelf life and improvement of the safety of the products are the key issues. High-pressure processing of fresh salmon and light preserved products like cold smoked salmon are reported; the effect on microorganisms, enzymes, and organoleptic properties are described. First, the general effects of high-pressure on food constituents mainly proteins are discussed. The secondary, tertiary and quaternary structure of proteins can unfold due to the effect of high pressure, resulting in the inactivation of enzymes. Enzymes contribute to the softening of salmon meat, off-taste and off-odors during storage. Lipids and proteins/enzymes are constitutes of the cell membranes. It is known that high-pressure can inactivate microorganisms by damaging their membranes. A further topic is related to the effect of high-pressure on the color. A bleaching of salmon meat after high-pressure processing is often observed. The color is mainly caused by the pigment astaxanthin. Due to the antioxidative properties of astaxanthin lipid oxidation and color should be related, as antioxidants can trap free radicals, preventing lipid oxidation. The process parameters, pressure, duration, and temperature affecting microorganisms, structure, and color are described in this chapter. All these effects have to be evaluated as they determine the market chances of the product.

Effects of High-Pressure on Food Constituents

The high-pressure treatment for the improvement of food shelf life has been known since 1889, when Hite applied this technology to inactivate bacteria in milk. Another pioneer of high pressure processing is Bridgman who denatured egg albumin proteins. Since then, high pressure processing of foods has been extensively studied during the last century but technological constraints for developing equipment have hindered economical commercialization of the process until the 1990s (Minerich, 2003).

Due to the availability of high-pressure technology nowadays and the growing consumer demand for valuable and fresh food high-pressure treatment becomes more important in the food industry. Moreover, high-pressure treatment is regarded as a "mild technology" because flavors and vitamins are maintained while enzymes and pathogenic microorganisms can be inactivated (Heinz and Knorr, 2002; Toepfl et al., 2006). Generally, high pressure processing, applied in food industries, goes up to 600 MPa. Such pressures do not affect covalent bonds of molecules. Phase transitions, alternations of the hydrogen bond network, and hydrophobic interactions of biomolecules are affected. Mainly weak intra- and intermolecular interactions related to the secondary, tertiary and quaternary structure of proteins are altered, including interactions between polar groups of the amino acid of the protein and bound water. Thus, most of the studies on seafoods were dedicated to the effect of high pressure treatment on fish proteins, inactivation of enzymes, and meat color. Another important issue is the effect of high-pressure treatment on lipids. Pure lipids undergo a phase transition by the application of high pressure, while pressurization of the whole food matrix often goes along with lipid oxidation, which causes rancidity and thus is also associated with spoilage and reduced shelf life, respectively. Both lipids and proteins/enzymes are the main constituents of membranes, high-pressure can therefore induce membrane perturbation, which is attributed to the promotion of phase transitions in the phospholipid bilayer from liquid to more tightly packed gel phases. High pressure processing can also denature or displace membrane-bound enzymes (Murchie et al., 2005). Moreover, alterations in morphology and internal organisation of cells, including cell lengthening, contraction of the cell wall and pore formation, separation of the cell membrane from the cell wall, and compression of gas vacuoles are affected by high-hydrostatic pressure (Cheftel, 1995). Finally, DNA and ribosome destruction have also been found in pressurized cells; a correlation between cell death and ribosome damage has been postulated (Murchie et al., 2005). These alterations and damages of the membranes and cells provide an explanation of the inactivation of microorganisms.

Microorganisms

Inactivation of various microorganisms is a fundamental challenge for the food industry and it is an actively developing field. In the so-called lightly preserved products like cold smoked salmon, Gram-negative spoilage flora is somewhat inhibited and shelf life is extended as compared to fresh iced fish. These products are not sterile and do allow microbial growth and typically spoil due to microbial activity in some time. Cold smoking of salmon causes a shift from a population dominated by Gram-negative bacteria in the raw product to Gram-positive organisms after processing. Generally species of spore-forming *Clostridium* and *Bacillus*, Gram-positive cocci and lactic acid bacteria are the dominant survivors. During the chilled storage these bacteria grow causing the spoilage (International Commission on Microbiological Specification for Foods, 2005).

High-pressure treatment at low temperatures combined with modified atmosphere packaging was used by Amanatidou et al. (2000) for the preservation of shelf life. According to their studies the initial flora present on fresh salmon was 10^4 CFU/g or higher and consisted mainly of pseudomonads. Yeasts, lactic acid bacteria, hydrogen sulfide-producing bacteria presumably *Shewanella putrefaciens* and Enterobacteriaceae were also present. High-pressure treatment at 150 MPa for 10 min did not reduce the initial flora significantly (< 1 log). Pressure treatment at 100 MPa for 30 min reduced the population of pseudomonads and *S. putrefaciens*, while the total flora was not significantly diminished. Lactic acid bacteria were completely inactivated after processing at 200 MPa independent of the pressure holding time. The critical level of microbial contamination 7.0 - 7.2 log CFU/g is a threshold value, which correlates with the development of fishy, amine-like off-odors. Modified atmosphere or a combination of high-pressure treatment followed by storage in 50% O_2 and 50% CO_2 atmosphere prolonged the critical level of microbial growth to 10-12 days, whereas vacuum-packed pressurized salmon reached this level already at 7-8 days. However, the high-pressure effect becomes evident by retardation of all bacterial groups to 18 days by modified package in 50% O_2 and 50% CO_2 atmosphere prior to high-pressure treatment.

Besides the saprophytes, pathogens were found in lightly preserved fish products, which have been implicated in outbreaks of food poisoning (International Commission on Microbiological Specification for Foods, 2005). In particular, *Listeria monocytogenes*, which can grow during chilled storage, is frequently detected on vacuum-packed cold smoked salmon. *L. monocytogenes* inoculated on salmon decreased by 1.5 log CFU after treatment at 150 MPa for 30 min of vacuum-packed samples. Packing in a 50% O_2 and 50% CO_2 atmosphere prior to high-pressure treatment led to a reduction of 3.5 log CFU. Another food-borne pathogen on salmon is *Salmonella typhimurium*. The same processing parameter as for *L. monocytogenes* causes a decrease of 1 and 2 log CFU, respectively. The reduction of the microbial growth of the high-pressure treatment in combination with the modified atmosphere is probably related to intracellular formation of reactive species as well as phase transition phenomena. Particularly, the latter are related to the properties of compressed CO_2, which is solved in the aqueous phase. Pressure release may lead to a sudden expansion of CO_2, causing cell damage and disruption of intracellular enzymes and organelles. As a model system (surrogate) the non-pathogenic *Listeria innocua* is often used instead of *L. monocytogenes* in food processing studies. Picart et al. (2005) found that *L. innocua* on smoked salmon mince were inactivated by pressurization at 207 MPa for 60 min by 1.1 log cycle at sub-zero temperatures (down to –21 °C) and to 0.7 log cycles at 4°C. Ice crystal formation was not observed at sub-zero temperatures. Whereas conventional freezing at –28 and –40 °C did not cause microbial inactivation, pressure shift freezing form 207 MPa at –22 °C to 0.1 MPa at –25 °C with pressure releasing times of approximately 3 s and 18 min induced a reduction of 1.4 and 1.9 log cycles, respectively. The highest decrease (2.5 log cycles) of *L. innocua* was obtained by a priori freezing at –28 °C for 24 h to the high pressure treatment at 207 MPa for 23 min (-29 °C) and a fast pressure release. The inactivation efficiency may be caused by a synergistic interaction of pressure and sub-zero temperature. The further reduction by a priory freezing may be related to successive water phase transitions including ice nucleation during decompression. The various high pressure-subzero temperature treatments did not indicate any significant sub-lethal injury of *L. innocua*. As the contamination level of *L. monocytogenes* is usually low, a reduction of 2 log cycles would improve the safety. In an other investigation pressurization at 250 MPa at 9 °C for 20 min did not inactivate *L. monocytogenes* (Lakshmanan and Dalgaard, 2004). If the behavior of the two bacterial species is similar, these data would support the synergistic effect of pressure and sub-zero temperature for the inactivation. However, a similar inactivation behavior of the high pressure treatments for *L. innocua* and *L. monocytogenes* has to be shown.

STRUCTURE

Smoking is one of the oldest food preservation approaches, enhancing texture and flavor of salmon. The cold-smoking process does not exceed 32 °C; therefore inactivation of enzymes is limited in the salmon tissue. Most of the deteriorative changes of vacuum-packed and chill-stored smoked salmon like tissue softening, off-flavors and off-odors are thus mainly caused by enzymes (Laksmanan et al., 2003). Among inherent enzymes, proteolytic enzymes, such as cathepsins and chaplains are the important ones. Cathepsin L and cathepsin B have been found to cause extensive softening of muscle in mature salmon. These enzymes are known to digest both myofibrillar proteins and connective tissues. Cathepsin L has also been associated with the production of jelly-like softening in some seafood and degradation of actomyosin into low molecular fragments at pH 5.5–7.0 as observed in mackerel. Lakshmanan et al. (2005) observed an activation of Cathepsin B-like and Cathepsin B + L-like at pressures around 100 MPa, while these enzymes were inactivated at pressures higher than 300 MPa. The activation at lower pressures is presumably assigned to the release from lysosomes (Homma et al., 1994). Calpain like the Cathepsin enzymes have the ability to hydrolyze muscle proteins at post-mortem pH. At 300 MPa, the activity of calpain enzymes was significantly decreased until 12 days of storage (Lakshmanan et al. 2005).

Besides the enzyme activity, physical properties related to the texture of meat like cutting strength and water holding capacity (WHC) of pressurized salmon were studied. The cutting strength of pressurized fresh salmon (150 MPa) did not show a significant effect compared to non-pressurized salmon, but an unacceptable increase of the hardness of vacuum packed salmon was found after 14 days (Amanatidou et al. 2000). The WHC describes the ability of muscles to keep the water. Therefore, the conformation of the muscle proteins is related to the WHC. Lakshmanan et al. (2007) found that pressure and processing time had significant effects on the moisture content on smoked salmon, so the WHC was increased by 2 % by 150 MPa and 10 min treatment, whereas both pressure and processing time did not affect fresh Atlantic salmon. The WHC were measured by the filter paper and spin-spin relaxation proton nuclear magnetic resonance.

Color Changes and Lipid Oxidation

It is known that pressurization causes undesired side effects including changes of the color of meat. Bleaching of red muscle meat from pork, beef, or tuna has been reported which is possibly due to denaturation of myoglobin (Cheftel and Culioli, 1997). Such changes have a negative impact on the consumer purchase decision of pressure-treated products since consumers judge the quality of food, *inter alia*, on the basis of the color. This is in particular true for vacuum-packed smoked salmon. Unlike to pork or tuna meat, heme proteins are not the main origin of the intense reddish color of wild salmon, which instead is primarily caused by the carotenoids, mainly astaxanthin (Figure 1), and to a smaller extent other carotenoids like canthaxanthin and β-carotene (Andersen et al., 1990). The color of carotenoids is caused by an allowed π-π* (S_0-S_2) transition in the visible region of the conjugated double bond system.

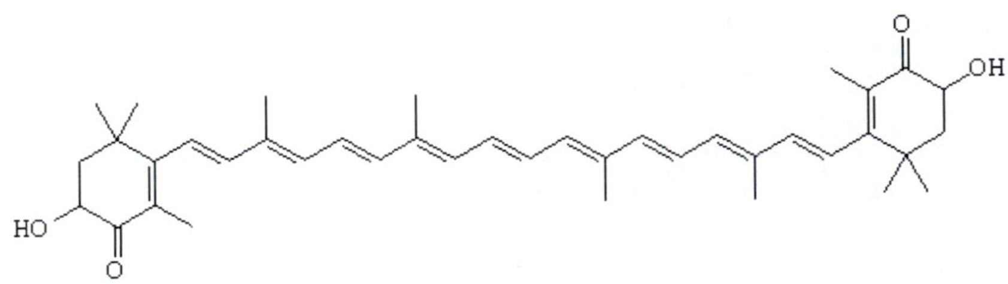

Figure 1. Stucture of astaxanthin

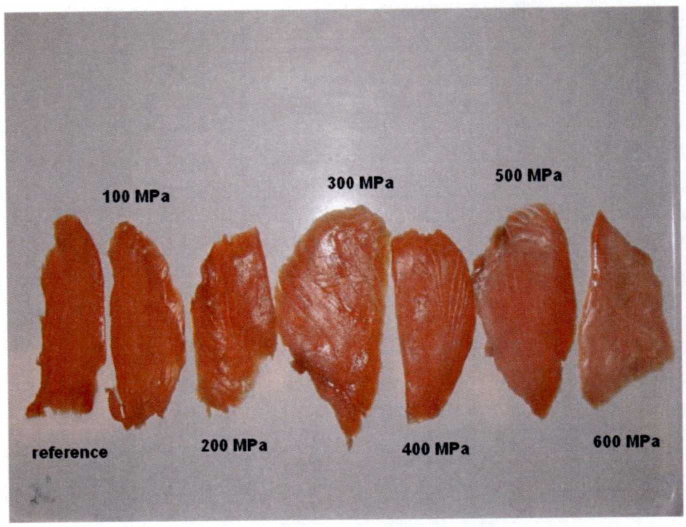

Figure 2. Smoked salmon after high-pressure treatment at room temperature for 5 min.

In salmon muscle astaxanthin is attached to proteins. Matthews et al. found, that α-actinin was the only myofibrillar protein correlating significantly with astaxanthin binding. However, their results suggest, that the astaxanthin binding protein, α-actinin, may not be the limiting factor for deposition of astaxanthin in the flesh of salmonids (Matthew et al., 2006).

Astaxanthin is a powerful antioxidant; it is several times more effective than other carotenoids or vitamin E (Guerin et al., 2003). Free radicals and highly reactive forms of oxygen can damage lipid membranes, which have a negative impact on human health. The antioxidative potential of astaxanthin might help to prevent and contribute to the healing of several human diseases.

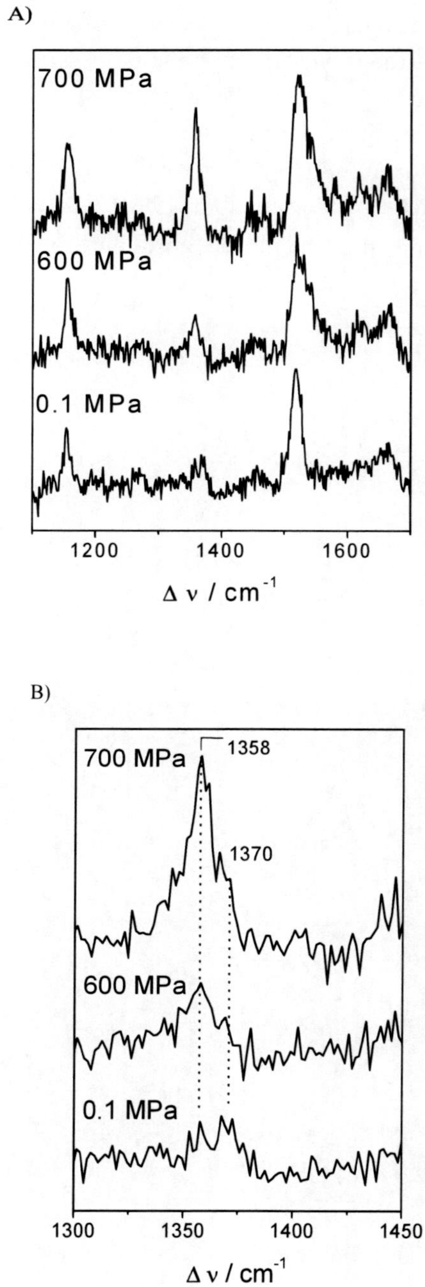

Figure 3. A) RR spectra of smoked salmon after pressurization (700 MPa and 600 MPa) and as a reference spectrum smoked salmon at ambient pressure (excitation wavelength 413 nm); B) enlargement of spectral region (Tintchev et al., 2009).

In a word, antioxidative activity of astaxanthin, lipid oxidation and color should be related in salmon meat, as astaxanthin can capture radicals and thus preventing lipid oxidation. In fact, it is known that carotenoids bleach i.e., lose their color, when exposed to radicals or to oxidizing species. This process involves degradation of the polyene system either by cleavage or by addition to one of the double bonds (Krinsky and Johnson, 2005).

High-pressure treated smoked salmon at room temperature for 5 min is shown in Figure 2. A bleaching is observed by increasing pressure. Currently, the processing parameters are optimized to minimize the loss of color and maintain the textural properties of the smoked salmon meat at the German Institute of Food Technologies. Short pressure holding times at low temperatures are promising process parameters for a successful application.

Reactive oxygen species degrade polyunsaturated lipids, forming malondialdehyde (MDA), which reacts with two equivalents of thiobarbituric acid (TBA) to give a fluorescent red derivative that can be assayed spectrophotometrically. Lakshmanan et al. (2005) found an increase of thiobarbituric acid reacting substances (TBARS) after high-pressure treatment time, i.e. high pressure (300 MPa) induces lipid oxidation. Pure lipids are not significantly affected by high-pressure treatment in the absence of oxygen. First, the presence of muscle tissue causes the increase in lipid oxidation (Cheftel and Culioli, 1997). It has been postulated that heme protein denaturation in muscle tissue caused by high-pressure induces lipid oxidation in fish muscle (Tanaka et al., 1991; Ohshima et al., 1992). Tintchev et al. (2009) indeed observed that pressure-treatment (600 and 700 MPa) facilitates the oxidative degradation of astaxanthin coupled to the reduction of metmyoglobin (met hemoglobin) direct in smoked salmon. Resonance Raman (RR) spectroscopy was applied in this study, which selectively probes the vibrational modes of chromophores or protein cofactors by excitation in resonance with an electronic transition. Therefore, this technique allows characterization of these molecules in complex matrices. Inasmuch as Raman spectroscopy is based on inelastic scattering, the sample has not to be translucent.

RR spectra of high-pressure processed smoked salmon are shown in Figure 3. Totally symmetric vibrational modes of the chromophore are dominant in the RR spectra. These are, specifically the C-C and C=C stretching modes of the polyene chain which are observed at 1155 cm^{-1} and 1520 cm^{-1}, respectively. The broadening of the latter by increasing pressure is accompanied by a slight frequency up-shift, possibly arising from the shortening of the conjugated double bond system and thus degradation of astaxanthin, providing an explanation for the observed bleaching of smoked salmon by high pressure processing.

Interesting is the band at ca. 1358 cm^{-1} which in the spectrum of the non-pressurized sample shows a very low intensity but a higher frequency at ca. 1370 cm^{-1}. High-pressure processing at 700 MPa for 10 min causes an increase of the former and a decrease of the latter. The 1358 and 1370 cm^{-1} bands cannot be assigned to astaxanthin but are at a position that is characteristic of the ν_4 mode of a ferrous and ferric heme protein, respectively (Oellerich et al., 2002). The most likely candidates are myoglobin or hemoglobin, which are major protein constituents of salmon flesh. This band is quite sensitive to the oxidation state of the heme protein. The increase of the band at 1358 cm^{-1} at the expense of the band at 1370 cm^{-1}, therefore, indicates a change in the oxidation state of the heme iron from Fe(III) to Fe(II). This process is obviously coupled to the degradation of astaxanthin. However, the mechanism of this process has still to be addressed.

CONCLUSION

Most of the cited work used pressure in the range of 300 MPa. Nowadays, commercial high-pressure plants can easily obtain 600 to 800 MPa. In this range inactivation of microorganisms is more effective (Rastogi et al., 2007). For instance, the shelf-life of a salmon spread could be extended from 60 to 180 days at 3 or 8 °C without significant chemical, microbiological, or sensory changes by high-pressure treatment at 700 MPa. By this pressure complete inactivation of pathogens present in the inoculated sample was obtained (Capri et al., 1995). Highly valuable foods, or highly heat sensitive products that can not be preserved by other techniques are suited for high-pressure treatment. The effects of high pressure on microorganisms and organoleptic properties are highly dependent on processing parameters and the complex nature of pressure necessitates the careful design of processing parameters to balance inactivation of enzymes and microorganisms, while maintaining optimal product quality. However, high-pressure processing offers advantages over conventional processing methods pertinent to seafood, as exemplified by the success of high-pressure treated oysters in the US. Besides the consumer benefit of fresh-like food, reduction of energy costs and its applicability to packaged foods are favorable.

As wild salmon belongs to highly valuable foods the cost of high-pressure processing can be afforded, which will add around 0.08 Euro per kg product by applying pressures at 600 MPa. EU legislation prescribes freezing of wild salmon for at least 24 h at –18 °C to assure elimination of the parasite, *Anisakis* spp., which does not inactivate during cold smoking. A high-pressure treatment should eliminate *Anisakis* spp. providing an alternative to freezing. Thus, high-pressure processing of fresh salmon should have good chances on the market.

References

Amanatidou, A; Schlüter, O; Lemkau, K; Gorris, LGM; Smid, EJ; Knorr, D. Effect of combined application of high pressure treatment and modified atmospheres on the shelf life of fresh Atlantic salmon, *Innovative Food Science & Emerging Technologies*, 2000, 1, 87-98.

Andersen, HJ; Bertelsen, G; Christophersen, AG; Ohlen, A; Skibsted, LH. Development of rancidity in salmonoid steaks during retail display, *Zeitschrift für Lebensmittel-Untersuchung und -Forschung A*, 1990, 191, 119-122.

Capri, G; Gola, S; Maggi, A; Rovere, P; Buzzoni, M. Microbial and chemical shelf-life of high-pressure treated salmon cream at refrigeration temperatures, *Industria Conserve*, 1995, 70, 386-397.

Cardinal, M; Knockaert, C; Torrissen, O; Sigurgisladottir, S; Moerkoere, T; Thomassen, M; Vallet, JL. Relation of smoking parameters to the yield, colour and sensory quality of smoked Atlantic salmon (Salmo salar), *Food Research International*, 2001, 34, 537-550.

Cheftel, JC; Culioli, J. *Effects of High Pressure on Meat: A Review*, *Meat Science*, 1997, 46(3), 211-236.

Cheftel, JC. Review: High pressure, microbial inactivation and food preservation, *Food Science and Technology International*, 1995, 1, 75-90.

Guerin, M; Huntley, ME; Olaizola, M. Haematococcus astaxanthin: application for human health and nutrition, *Trends in Biotechnology*, 2003, 21, 210-216.

Heinz, V; Knorr, D. Effects of high pressure on spores. Ultra high pressure treatments of foods, In MEG. Hendrickx & Knorr (Eds.), *Ultra high pressure treatments on foods* (77-114) Kluwer Academic/ Plenum Publishers, New York, 2002.

Homma, N; Ikechi, Y; Suzuki, A. *Effects of high-pressure treatment on the proteolytic enzymes in meat Meat Science*, 1994, 38, 219-228.

International Commission on Microbiological Specification for Foods Microorganisms in foods 6, *Microbial ecology of food commodities, second edition*, Kluwer Academic/Plenum Publishers, New York, 2005.

Krinsky, NI; Johnson, EJ. Carotenoid actions and their relation, *Molecular Aspects of Medicine*, 2005, 26, 459-516.

Lakshmanan, R; Dalgaard, P. Effect of high pressure processing on Listeria monocytogenes, spoilage microflora and multiple compound quality indices in chilled cold-smoked salmon, *Journal of Applied Microbiology*, 2004, 96, 398-408.

Lakshmanan, R; Miskin, D; Piggott, JR. Quality of vacuum packed cold-smoked salmon during refrigerated storage as affected by high-pressure processing, *Journal of the Science of Food and Agriculture*, 2005, 85, 655-661.

Lakshmanan, R; Parkinson, JA; Piggott, JR. *High-pressure processing and water-holding capacity of fresh and cold-smoked salmon (Salmo salar)*, LWT 2007, 40, 544-551.

Lakshmanan, R; Patterson, MF; Piggott, JR. Effects of high-pressure processing on proteolytic enzymes and proteins in cold-smoked salmon during refrigerated storage, *Food Chemistry*, 2005, 90, 541-548.

Lakshmanan, R; Piggott, JR., Paterson, A. Potential applications of high pressure for improvement in salmon quality, *Trends in Food Science & Technology*, 2003, 14, 354-363.

Matthews, SJ; Ross, NW; Lall, SP; Gill, TA. Astaxanthin binding protein in Atlantic salmon, *Comparative Biochemistry and Physiology B: Biochemistry and Molecular Biology*, 2006, 144, 206-214.

Minerich, PL; Labuza, TP. Development of a pressure indicator for high hydrostatic pressure processing of foods, *Innovative Food Science & Emerging Technologies*, 2003, 4, 235-243.

Murchie, LW; Cruz-Romero, M; Kerry, JP; Linton, M; Patterson, MF; Smiddy, M; Kelly, AL. High pressure processing of shellfish: A review of microbial and other quality aspects, *Innovative Food Science & Emerging Technologies*, 2005, 6, 257-270.

Oellerich, S; Wackerbarth, H; Hildebrandt, P. Spectroscopic characterization of nonnative conformational states of cytochrome c, *Journal of Physical Chemistry*, B 2002, 106, 6566-6580.

Ohshima, T; Nakagawa, T; Koizumi, C. Effect of high hydrostatic pressure on the enzymatic degradation of phospholipids in fish muscle during storage, in *Seafood Science and Technology*, ed by Bligh EG, Fishing News Book, Oxford, 1992, 64-75.

Picart, L; Dumay, E; Guiraud, JP; Cheftel, JC. Combined high pressure-sub-zero temperature processing of smoked salmon mince: phase transition phenomena and inactivation of Listeria innocua, *Journal of Food Engineering*, 2005, 68, 43-56.

Rastogi, NK; Raghavarao, KSMS; Balasubramaniam, VM; Niranjan, K; Knorr, D. Opportunities and Challenges in High Processing of Foods, *Critical Reviews in Food Science and Nutrition*, 2007, 47, 69-112.

Tanaka, M; Xueyi, Z; Nagashima, Y; Taguchi, T. *Effect of high pressure on lipid oxidation in sardine meat Nippon Suisan Gakkaishi*, 1991, 57, 957-963.

Tintchev, F; Kuhlmann, U; Wackerbarth, H; Töpfl, S; Heinz, V; Knorr, D; Hildebrand, P. Redox Processes in Pressurized Smoked Salmon Studied by Resonance Raman Spectroscopy, *Food Chemistry*, 2009, 112, 482-486.

Töpfl, S; Mathys, A; Heinz, V; Knorr, D. Review: Potential of emerging technologies for energy efficient and environmentally friendly food processing, *Food Reviews International*, 2006, 22, 405-423.

In: Salmon: Biology, Nutrition and Consumption
Editors: P. Lacopo et al., pp. 169-186
ISBN: 978-1-60741-287-8

Chapter 7

WELFARE EFFECTS OF SUPPLY EXPANSION WITH TRADE RESTRICTIONS: THE CASE OF SALMON*

Henry W. Kinnucan[#] and Øystein Myrland

Department of Agricultural Economics and Rural Sociology, Auburn University, USA
Department of Economics and Management, University of Tromsø, Norway

ABSTRACT

In response to protectionist pressures from United Kingdom salmon producers, in 1996 Norway instituted a feed quota to limit its production. The feed quota made the Norwegian and world supply curves for salmon less price elastic. This contributed to price instability, magnified the downward pressure on world salmon prices associated with the 71% increase in supplies of farmed salmon from Chile between 2000 and 2002, and increased Norway's incidence of US salmon tariffs. Our analysis suggests these unintended consequences of the feed quota policy are nontrivial, amounting to some $62 million in lost surplus to UK producers between 2000 and 2002, equivalent to 9.5% of export value.

INTRODUCTION

Traditionally, Norway has been the world's largest producer of salmon, accounting for some 46% of total world production in 1995 (Asche 1997). However, Norway's dominant position is being challenged by emerging low-cost producers, most notably Chile. In 2002

---

* Final revision submitted to *Journal of International Agricultural Trade and Development* 24 September 2005.

Kinnucan is a professor in the Department of Agricultural Economics and Rural Sociology, Auburn University, USA and Myrland is an associate professor in the Department of Economics and Management, University of Tromsø, Norway. Appreciation is expressed to Tom Sebulonsen and Jan Trollvik of the Norwegian Seafood Export Council for providing data and funding and to two anonymous journal reviewers for their helpful comments. An earlier version of this paper was presented at the 50th Anniversary Workshop on Global Aquaculture sponsored by the Centre for Fisheries Economics, Bergen, Norway, 11 June 2003. Responsibility for final content, however, rests strictly with authors.

Chile accounted for 25% of the world's farmed salmon production, up from 20% in 1995. Perhaps most telling is that between 2000 and 2002 the ratio of Chile's price to Norway's price declined from 0.99 to 0.74 (table 1).

Chile's growing price advantage presents a challenge to the wild-caught salmon sector, which is located chiefly in North America. But it also poses a challenge to European producers of farmed salmon, especially producers in the United Kingdom where production costs may be higher than in Norway.[1] It should not be surprising, therefore, that producers in the UK and US have sought protection through tariffs and other measures designed to reduce the supply of Norwegian and Chilean salmon on world markets. In particular, in response to pressures from UK producers, in 1996 Norway instituted a feed quota in an effort to reduce supply and strengthen market prices.

Table 1. Baseline Data for the World Salmon Model

Variable	Definition	Value			
		2000	2001	2002	Average
P_N	Norway price	3.78	2.77	3.07	3.21
P_{CH}	Chile price	3.76	2.68	2.26	2.90
P_{UK}	UK price	4.06	2.96	3.25	3.42
P_{CAN}	Canada price	5.61	5.24	4.81	5.22
P_{FI}	Faroe Islands' price	3.73	2.28	2.55	2.85
P_{RS}	ROW price (producing countries)	2.93	3.18	3.43	3.18
P_{EU}	EU price	3.93	2.77	3.12	3.27
P_{US}	US price	5.22	4.67	4.36	4.75
P_J	Japan price	6.23	4.76	5.33	5.44
P_{RD}	ROW price (consuming countries)	4.98	3.40	3.37	3.92
X_N	Norway's net exports	409	399	438	415
X_{CH}	Chile's net exports	138	203	236	192
Variable	Definition	Value			
		2000	2001	2002	Average
X_{CAN}	Canada's net exports	46	56	65	56
X_{FI}	Faroe Islands' net exports	31	44	42	39
X_R	ROW's net exports	13	19	18	17
ΣX_i	*Total net exports*	*700*	*788*	*859*	*782*
M_{EU}	EU's net imports	392	409	410	404
M_{US}	US' net imports	179	224	259	221
M_J	Japan's net imports	39	43	38	40
M_R	ROW's net imports	90	112	152	118
ΣM_j	*Total net imports*	*700*	*788*	*859*	*782*

Source: Norwegian Seafood Export Council and government statistics.

Note: Prices are fob expressed in constant US 2001 dollars per kilogram; quantities are raw weight expressed in 1000 metric tons.

[1] Although Scottish producers claim their production costs are higher than in Norway due to limits on fish farm sites and extra tax burdens, no data exist to support the claim. A study has been commissioned to determine the comparative costs of salmon farming in different countries (Callender McDowell 2005), which should help to clarify the issue.

About the same time the United States imposed countervailing and dumping duties on Norwegian salmon averaging 26.1%, which were followed by similar duties on Chilean salmon averaging 5.5%.

More recently, the European Commission has imposed minimum import prices and a tariff-rate quota on farmed salmon imports and a provisional anti-dumping duty of between 6.8% and 24.5% on imports from Norway in an effort to protect Scottish and Irish producers (Bendz 2005; Lem 2005).

Ordinarily supply control is used to protect *domestic* producers by taking advantage of an inelastic demand (Hertel and Tsigas 1991; Kola 1993; Edelman, Langley and White 2003). Norway's feed quota is unique in that it is designed to protect *foreign* (namely UK) producers by limiting the supply of Norwegian salmon in the European Union market. Because most studies suggest the export demand for salmon is price elastic (Bjørndal, Salvanes, and Andreassen 1992; Hermann, Mittlehammer and Lin 1993; Asche, Bjørndal and Salvanes 1998), and Norway exports over 90% of its production, the supply restriction is not likely to benefit Norway's producers from a revenue perspective. At issue here, however, are the more subtle impacts stemming from policy interactions and induced supply growth.

The purpose of this research is to determine the unintended welfare losses from Norway's feed quota policy. As noted by Houck (1986, p. 181) "most...trade policy schemes accentuate the price inelasticity of world supply and demand functions." In the case of the feed quota, we show that the inelasticity introduced into the world supply curve for salmon aggravated price instability, magnified period-to-period swings in trade volumes, undermined US tariff policy, and intensified producer losses associated with increased supplies of farmed salmon from Chile. The feed quota also diluted the spillover benefits that UK and other international competitors enjoy as a result of the US tariffs targeted against Norway and Chile.

The analysis proceeds by first specifying a partial equilibrium model of the world salmon market similar to the one used by Kinnucan and Myrland (2000, 2002) to assess the Norway-EU salmon agreement. The model is then simulated to determine the price, trade, and welfare effects of Chile's supply expansion and US tariffs with and without the feed quota. A concluding section summarizes the key findings.

MODEL

Specification

The model is based on two simplifying assumptions: the world salmon market is sufficiently competitive that the Law of One Price (LOP) holds, and different salmon species (e.g., Pacific and Atlantic) are sufficiently close substitutes that the generalized Composite Commodity Theorem holds. Empirical support for LOP for *long-run* analysis (the major focus of this paper) is provided in the studies by DeVoretz and Salvanes (1993), Asche, Bremmes, and Wessels (1999), Steen and Salvanes (1999), and Asche (2001). As for whether it is valid to treat different salmon species as a single good as implied by CCT, Asche, Bremmes, and Wessels (1999, p. 579) state "when investigating long-run market issues ...

studies that aggregate salmon from species-specific to a generic product category are capturing much of the information about the international market."

With the foregoing assumptions, the supply side of the model consists of six equations as follows:

$$X_N^* = \acute{\varepsilon}_N P_N^* \text{ (with feed quota)} \quad (1a)$$

$$XN^* = \varepsilon N\, PN^* \text{ (without feed quota)} \quad (1b)$$

$$XCH^* = \varepsilon CH\, (PCH^* - \gamma CH) \quad (2)$$

$$Xi^* = \varepsilon i\, Pi^* \quad (3) - (6)$$

where equations (1) and (2) refer to export supplies from Norway and Chile, respectively, and equations (3) - (6) refer to export supplies from UK, Canada, Faroe Islands, and Rest-of-World (ROW). (For variable definitions and numerical values, see table 1.) The asterisks (*) in these equations denote percentage changes (e.g., $X_i^* = dX_i/X$), the ε_i refer to excess supply elasticities, and the γ_{CH} parameter in equation (2) refers to the vertical shift in Chile's excess supply curve over the study period (to be discussed later). The excess supply curve for Norway is specified in two equations to accommodate the supply elasticity reducing aspect of the feed quota wherein $\acute{\varepsilon}_N < \varepsilon_N$. The disaggregation of supply into six regions permits a detailed assessment of the unintended consequences of the feed quota.

The demand side of the market is separated into four regions as follows:

$$M_j^* = - \eta_j P_j^* \quad (7) - (10)$$

where the j subscript refers to the EU, US, Japan, and ROW. The η_j parameters represent excess demand elasticities, which are expressed in absolute value. The EU dominates among the customer markets for salmon with a trade share of 52%, followed by the US, ROW and Japan with shares of 28%, 15%, and 5% respectively as shown in table 2.

In the early 1990s the US imposed tariffs on imports from Norway averaging 26.1%, which were later followed by tariffs on imports from Chile averaging 5.5%. These tariffs are modeled using the following identity implied by LOP:

$$PUS = (Pk + Ck)(1 + \tau k)$$

where $k = N, CH$ denotes Norway and Chile, P_{US} is the US price of salmon inclusive of transportation costs and the tariff, P_k is price in the named country exclusive of transportation costs and the tariff, C_k is the per-unit transportation cost, and τ_k is the *ad valorem* tariff.

Taking the logarithmic differential of this equation holding C_k constant yields:

d ln PUS = (Pk/PUS)(1 + τk) d ln Pk + (τk/(1 + τk)) d ln τk,

which may be written more compactly as:

PUS* = (1 + τk) φk Pk* + ζk τk*

where $\varphi_k = P_k/P_{US}$ are price transmission elasticities and $\zeta_k = \tau_k/(1 + \tau_k)$ are tariff transmission elasticities. If tariffs in the initial equilibrium are zero, $\tau_k = 0$ and the above equation simplifies to:

PUS* = φk Pk* + τk′ *(11) - (12)*

where τ_k' are the tariff rates in post-tariff equilibrium. Setting $\tau_N' = 0.261$ and $\tau_{CH}' = 0.055$, equations (11) and (12) imply the US price will rise by 26% in response to the Norwegian tariff and 5.5% in response to the Chilean tariff *provided the tariffs have no effect on prices in the targeted countries.* In reality, the tariffs will cause prices in the targeted countries to decline as import demand from the US decreases. Hence, the US price is expected to rise by less than the amount of the tariffs. A critical issue in this research is the extent to which the feed quota reduces the ability of US tariffs to raise the US price.[2]

Table 2. Mean Values for Quantity Share Parameters, 2000-2002

Item	Definition	Value[a]
kx_N	Norway's share of world exports ($X_N/\Sigma X_i$)	0.531
kx_{CH}	Chile's share of world exports ($X_{CHL}/\Sigma X_i$)	0.246
kx_{UK}	Scotland's share of world exports ($X_{UK}/\Sigma X_i$)	0.081
kx_{CAN}	Canada's share of world exports ($X_{CAN}/\Sigma X_i$)	0.071
kx_{FI}	Faroe Islands' share of world exports ($X_{FI}/\Sigma X_i$)	0.050
kx_R	ROW's share of world exports ($X_R/\Sigma X_i$)	0.021
km_{EU}	European Union's share of world imports ($M_{EU}/\Sigma M_j$)	0.516
km_{US}	United States' share of world imports ($M_{US}/\Sigma M_j$)	0.282
km_J	Japan's share of world imports ($M_J/\Sigma M_j$)	0.051
km_R	ROW's share or world imports ($M_R/\Sigma M_j$)	0.151
x_N	Share of Norway's production exported (X_N/S_N)	0.97
x_{CH}	Share of Chile's production exported (X_{CH}/S_{CH})	0.95
x_{UK}	Share of Scotland's production exported (X_{UK}/S_{UK})	0.53

[2] The price wedge associated with a tariff equals the tariff only if the price transmission elasticity is unity. This may be seen by setting $P_{US}* = 0$ in equations (11) and (12) and solving for foreign price to yield $P_k* = -\tau_k'/\varphi_k$. In this equation P_k* is the maximal decline in foreign price, i.e., the decline when the tariff's entire incidence is shifted to foreign producers. As can be seen, $P_k* = -\tau_k'$ only if $\varphi_k = 1$. If $\varphi_k < 1$, as is true in this study, it is possible for the price declines in Norway and Chile to exceed their respective tariffs, a fact to bear in mind when interpreting results.

Table 2 (Continued)

x_{CAN}	Share of Canada's production exported (X_{CAN}/S_{CAN})	0.90
x_{FI}	Share of Faroe Islands' production exported (X_{FI}/S_{FI})	0.99
x_R	Share of ROW's production exported (X_R/S_R)	0.60
m_{EU}	Share of EU's consumption imported (M_{EU}/D_{EU})	0.96
m_{US}	Share of US's consumption imported (M_{US}/D_{US})	0.93
m_J	Share of Japan's consumption imported (M_J/D_J)	0.59
m_R	Share of ROW's consumption imported (M_R/D_R)	0.90

Source: Norwegian Seafood Export Council.

Equations (11) and (12) link the US price to prices in Norway and Chile. Price linkages to the remaining countries are given by:

Table 3. Parameter Definitions and Values

Parameter	Definition	Value
$\acute{\varepsilon}_N$	Norway's excess supply elasticity with feed quota	0.44
ε_N	Norway's excess supply elasticity without feed quota	1.60
ε_{CH}	Chile's excess supply elasticity	1.64
ε_{UK}	Scotland's excess supply elasticity	3.89
ε_{CAN}	Canada's excess supply elasticity	1.80
ε_{FI}	Faroe Islands' excess supply elasticity	1.53
ε_R	Rest-of-world's excess supply elasticity	3.30
γ_{CH}	Vertical shift in Chile's excess supply curve between 2000 and 2002[a]	-0.58
η_{EU}	European Union's excess demand elasticity	1.31
η_{US}	United States' excess demand elasticity	1.40
η_J	Japan's excess demand elasticity	3.08
η_R	Rest-of-world's excess demand elasticity	1.50
φ_N	US-Norway price transmission elasticity	0.67
φ_{CH}	US-Chile price transmission elasticity	0.61
φ_{UK}	US-UK price transmission elasticity	0.72
φ_{CAN}	US-Canada price transmission elasticity	1.10
φ_{FI}	US-Faroe Islands' price transmission elasticity	0.60
φ_{RS}	US-ROW exporters' price transmission elasticity	0.67
φ_{EU}	US-EU price transmission elasticity	0.69
φ_J	US-Japan price transmission elasticity	1.14
φ_{RD}	US-ROW importers' price transmission elasticity	0.82
τ_N'	Norway's average tariff rate	0.261
τ_{CH}'	Chile's average tariff rate	0.055

[a] Computed using text equation (22); see narrative for details.

$$PUS^* = \varphi l \; Pl^* \qquad (13) - (19)$$

where l indexes countries other than Norway, Chile and US and $\varphi_l = P_l/P_{US}$ are price transmission elasticities computed in an analogous fashion as for equations (11) and (12). (See table 3 for numerical values.) The model is closed by the following market-clearing condition:

$$\Sigma_i^6 \, kx_i \, X_i^* = \Sigma_j^4 \, km_j \, M_i^* \qquad (20)$$

where kx_i and km_j are export quantity and import quantity shares, respectively, as given in table 2.[3] Equilibrium, therefore, requires that the changes in exports weighted by their respect export shares equal changes in imports weighted by their respective import shares.

Altogether the model contains 20 endogenous variables: ten to represent region-specific changes in prices, six to represent region-specific changes in exports, and four to represent region-specific changes in imports. There are three exogenous variables, two to indicate price wedges associated with US tariffs (the τ_k' parameters in equations (11) and (12)), and one to indicate Chile's supply expansion (the γ_{CH} parameter in equation (2)).

Parameterization

Numerical values for the excess supply and demand elasticities in the model were developed using the following formulas:

$$\varepsilon i = (eS + (1 - xi)\, eD)/xi \qquad (21a)$$

$$\eta j = ((1 - mj)\, eS + eD)/mj \qquad (21b)$$

where x_i are domestic export shares for the six exporting regions and m_j are domestic import shares for the four importing regions as defined in table 1.[4] The e_S and e_D parameters in equation (21) are domestic supply and demand elasticities for salmon within the specific exporting and importing regions. Given the dearth of region-specific estimates for these parameters, they are assumed to be uniform. Specifically, we set e_S to 1.5 in all regions except Norway, where e_S is set to 0.39 to reflect the elasticity-reducing aspect of the feed quota quantified by Kinnucan and Myrland (2002, p. 219). The value of 1.5 for e_S is based on Steen, Asche, and Salvanes's (1997) estimate of Norway's long-run supply elasticity prior to the feed quota, the only known empirical estimate of domestic supply response for salmon. Similarly, we set e_D to 1.2 in all regions, a value suggested by Bjørndal, Asche, and Steen's (1996) literature review.[5]

The foregoing procedure resulted in excess supply elasticities ranging from 0.44 for Norway to 3.89 for the UK and excess demand elasticities ranging from 1.31 for the EU to 3.08 for Japan (see table 3 for complete listing). The trade share weighted average excess demand elasticity is 1.45, which suggests increases in world exports can be absorbed without reductions in export value, an hypothesis to be examined later.

---

[3] The export and import data used in this analysis are *net* in that they take into account exporting countries' imports and importing countries' exports. For brevity, in the narrative we omit the "net" qualifier.

[4] Houck (1986, pp. 33-34) derives the same equations in a somewhat different form.

[5] Given the dynamic nature of salmon markets in recent years, a reviewer questioned whether the domestic demand elasticity estimate of 1.2 is still valid. For example, a recent study by Fousekis and Revell (2004) suggests the demand for salmon in Great Britain is price inelastic, a result consistent with supply increasing along a static linear demand curve. Whether this result can be generalized to other customer markets is an issue for further research. Sensitivity analysis with e_D set to 0.5 indicated little effect on results reported later.

According to data given in table 1 Chile's exports between 2000 and 2002 increased by 71% and Chile's price declined by 40%. These price and quantity changes were converted to the implied shift in Chile's excess supply curve using the following approximation formula (see Appendix for derivation):

$$\gamma CH \approx (\varepsilon CH\ PCH^* - XCH^*)/(\varepsilon CH + XCH^*) \qquad (22)$$

where γ_{CH} indicates the *vertical* shift, i.e., the shift in the price direction holding quantity constant. Substituting $X_{CH}^* = 0.71$, $P_{CH}^* = -0.40$ and and $\varepsilon_{CH} = 1.64$ into expression (22) yields $\gamma_{CH} = -0.581$. Hence, our analysis is based on the assumption that Chile's excess supply curve between 2000 and 2002 shifted down by 58.1%.[6] Since the shift is measured in the price direction, the 58.1% may be interpreted as the estimated reduction in Chile's per-unit production costs over the period in question. By way of comparison, Anderson (2002, p. 145) reports cost reductions of some 45% in the early years of salmon aquaculture development in Norway. At issue is the extent to which the associated supply expansion in Chile harmed competing exporters, and the extent to which the feed quota magnified the harm, especially with respect to UK producers.

A caveat in using expression (22) is that it is an approximation formula. To minimize approximation error Piggott (1992, p. 133) suggests limiting displacements to about 10%. However, in Marsh's (2003) analysis where a similar formula is used a 66% displacement was modeled with no apparent ill effects. Still, the potential for approximation error must be borne in mind when interpreting results.

PRICE AND TRADE EFFECTS

Houck (1986, p. 181) asserts that the inelasticity introduced into world supply or demand curves due to trade restrictions exacerbates price instability, but also magnifies fluctuations in trade volumes. To test this, and to provide a basis for welfare measurement, the price and trade effects of the US tariffs and Chile's supply expansion with and without Norway's feed quota are computed by expressing the model in matrix notation as follows:

$$A\ y = B\ x, \qquad (23)$$

where $\boldsymbol{A}$ is a 20 x 20 matrix of parameters corresponding to the model's endogenous variables; $\boldsymbol{y}$ is a 20 x 1 vector containing the model's endogenous variables; $\boldsymbol{B}$ is a 20 x 3 matrix of parameters corresponding to the model's exogenous variables; and $\boldsymbol{x}$ is a 3 x 1 vector containing the model's exogenous variables. Inverting A and pre-multiplying both sides of equation (23) by A^{-1} yields:

---

[6] The implied horizontal shift, i.e., the shift in the quantity direction with price held constant, is 95%. The price-constant supply increase of 95% is larger than the actual supply increase of 71% due to the depressing effects of the 40% price decline on Chile's production.

$$y = E\,x \qquad (24)$$

where $\boldsymbol{E} = \boldsymbol{A}^{-1}\boldsymbol{B}$ is a 20 x 3 matrix containing the desired price and trade effects. In computing $\boldsymbol{E}$ we set ε_N alternatively to 0.44 and 1.60 to assess the impact of the feed quota. For brevity, the price and trade effects for the Chilean tariff are not reported as they are qualitatively similar to those of the Norwegian tariff.

Results support the hypothesis that the feed quota contributes to instability in both prices and trade volumes (table 4).

Specifically, without the feed quota Chile's supply expansion would have caused export prices worldwide to decline by 7.7% and import prices to decline by 6.5%; with the quota the estimated declines are 9.7% and 8.2% respectively. Similarly, without the feed quota Chile's supply expansion would have caused exports and imports worldwide to increase by 9.3%; with the quota the estimated increase is 11.7%. Overall, these estimates suggests the feed quota enlarged price and trade volume fluctuations by 26%.

Table 4. Simulated Price and Quantity Effects of Chile's Supply Expansion and US Tariffs on Salmon Imports from Norway with and without Norway's Feed Quota

Endogenous Variable	Supply Expansion Effect		Ratio	Tariff Effect		Ratio
	With Feed Quota (%)	Without Feed Quota (%)		With Feed Quota (%)	Without Feed Quota (%)	
P_N*	-9.7	-7.8	1.26	-35.2	-28.0	1.26
P_{CH}*	-10.7	-8.5	1.26	4.2	12.1	0.35
P_{UK}*	-9.1	-7.2	1.26	3.5	10.2	0.35
P_{CAN}*	-5.9	-4.7	1.26	2.3	6.7	0.35
P_{FI}*	-10.9	-8.7	1.26	4.2	12.2	0.35
$\Sigma_{i=1}^{6}$ kx_i P_i*	*-9.7*	*-7.7*	*1.26*	*-16.9*	*-9.7*	*1.73*
P_{EU*}	-9.5	-7.5	1.26	3.7	10.7	0.35
P_{US}*	-6.5	-5.2	1.26	2.5	7.4	0.35
P_J*	-5.7	-4.5	1.26	2.2	6.4	0.35
$\Sigma_{j=1}^{4}$ km_j P_j*	*-8.2*	*-6.5*	*1.26*	*3.2*	*9.3*	*0.35*
X_N*	-4.3	-12.4	0.35	-15.5	-44.8	0.35
X_{CH}*	77.6	81.1	0.96	6.8	19.8	0.35
X_{UK}*	-35.2	-28.1	1.26	13.7	39.7	0.35
X_{CAN}*	-10.7	-8.5	1.26	4.2	12.0	0.35
X_{FI}*	-16.6	-13.2	1.26	6.5	18.7	0.35
$\Sigma_{i=1}^{6}$ kx_i X_i*	*11.7*	*9.3*	*1.26*	*-4.5*	*-13.1*	*0.35*
M_{EU}*	12.4	9.9	1.26	-4.8	-14.0	0.35
M_{US}*	9.1	7.3	1.26	-3.6	-10.3	0.35
M_J*	17.6	14.0	1.26	-6.8	-19.8	0.35
$\Sigma_{i=1}^{4}$ km_j M_j*	*11.7*	*9.3*	*1.26*	*-4.5*	*-13.1*	*0.35*

Note: to conserve space price and quantity effects for ROW are not presented.

Results also support the hypothesis that the feed quota undermines US tariff policy. In particular, without the feed quota the rise in US price associated with the 26.1% Norwegian tariff is 7.4%; with the quota the price rise dwindles to 2.5%. The reason for the tiny price effect is that most of the tariff's incidence is borne by Norwegian producers. Specifically, the decline in the Norwegian price associated with the tariff increases from 28.0% to 35.2% as the feed quota is applied. Hence, the feed quota shifts more of the tariff's incidence onto Norwegian producers, but also widens the price wedge (from 35.4% to 37.7%).[7] These estimates are consistent with the analyses by Asche (2001) and by Kinnucan (2003), which suggest anti-dumping tariffs do more to punish producers in the targeted exporting countries than to assist producers in importing countries.

Tariffs provide an implicit subsidy to exporters not subject to the tariff. However, in the present case the implicit subsidies are modest, with no single exporter enjoying a price rise greater than 4.2% due to the Norwegian tariff (table 4). (For the 5.5% Chilean tariff, the largest price rise enjoyed by Chile's international competitors is 1.7%.) Still, removal of the feed quota would increase the implicit subsidy by a factor of three; hence, foregone spillover benefits for UK producers in particular are not inconsequential.

Overall, Chilean supply expansion reduced export prices worldwide by 9.7% and increased imports worldwide by 11.7% (table 4). This implies that the "total" import demand curve (Pearce 1952; Buse 1958) for salmon is elastic at -1.2. Hence, supply control measures that reduce exports are counterproductive in that they erode export value. Indeed, the added supplies from Chile over the period in question *raised* export value by 2%.

WELFARE EFFECTS

To place welfare measures on the unintended consequences, we used the following formulas:

Producer Impacts:

$$\Delta PSCH = PCH\ XCH\ (PCH^* - \gamma CH)(1 + \tfrac{1}{2}\ XCHi^*) \qquad (25a)$$

$$\Delta PS_i = P_i\ X_i\ P_i^*\ (1 + \tfrac{1}{2}\ X_i^*) \quad i = N, UK, CAN, FI, ROW \qquad (25b)$$

Consumer impacts:

$$\Delta CSj = -\ Pj\ Mj\ Pj^*\ (1 + \tfrac{1}{2}\ Mj^*) j = EU, US, J, ROW \qquad (26)$$

Total impact:

[7] The widening of the price wedge is due to the non-unitary price transmission elasticities (see footnote 2). If all price transmission elasticities in the model are set to one, the simulated price wedges equal the respective tariffs, and are unaffected by the feed quota. Incidence, however, is affected in the manner indicated.

$$\Delta TS = \Sigma_{i=1}^{6} \Delta PS_i + \Sigma_{j=1}^{4} \Delta CS_j \qquad (27)$$

where ΔPS_i is the change in "producer" surplus in the ith exporting country, ΔCS_j is the change in "consumer" surplus in the jth importing country; and ΔTS is the change in total surplus.[8] In these equations P_i and P_j are average export and import prices, respectively, over the evaluation period as reported in table 1, and X_i and M_j are the corresponding export and import quantities.[9] The asterisked variables in equations (25) - (27) indicate the price and quantity impacts of the exogenous variables computed from equation (24). Since an increase in Chile's supply reduces the demand for salmon from competing exporters, equation (25b) measures competing supplier impacts by shifting the respective excess demand curves along each exporter's stationary excess supply curve. Conversely, the own impact of supply expansion is measured via equation (25a) by shifting Chile's excess supply curve along its stationary excess demand curve. (The term $(P_{CH}^* - \gamma_{CH}) > 0$ in equation (25a) indicates the extent to which the cost reduction associated with the supply shift exceeds the price reduction.) Consumer gains are measured via equation (26) by shifting the excess supply curves facing the importing countries along their respective stationary excess demand curves. All gains and losses are reported in constant (2001) US dollars.

Supply Expansion Effects

Chile's supply expansion generates a total surplus gain of $1.30 billion (table 5). Most of the gain ($1.05 billion) accrues to Chilean producers, as might be expected since the implied cost reduction (58.1%) far exceeds the associated price decline (10.7%, see table 4). With lower prices consumers gain $771 million. The losers, of course, are Chile's international competitors, who suffer a combined loss of $525 million. Most of this loss, $381 million, is absorbed by Norway's producers, with Canadian and UK producers a distant second at $49 million apiece.

Removal of the feed quota leaves the total welfare gain unchanged at approximately $1.30 billion, but shifts the incidence in favor of producers. Specifically, with the feed quota in place the producer share of the $1.3 billion total welfare gain is 41%; with the feed quota removed the producer share of the gain rises to 54%. The reduced consumer incidence may be explained by the fact that quota removal makes supply more elastic, which reduces the price decline associated with Chile's supply expansion. With price reduction attenuated, consumers gain less, and Chile's international competitors lose less, both of which help to tilt total welfare gain in favor of producers. Overall, estimates in table 5 suggest the feed quota enlarged benefits to consumers from the supply shift by 28%, but at a cost in increased losses to producers (exclusive of Chile) of between 20% and 31%. Losses to UK producers, the major focus of this analysis, decline from $49 million to $40 million with quota removal.

---

[8] Technically, surplus changes are measured off excess supply and demand curves and thus the "producer" and "consumer" surpluses are properly interpreted as "exporter" and "importer" surpluses. However, since trade accounts for the bulk of production or consumption in most of the regions (see table 2), the distinction in the present case is unimportant.

[9] A reviewer questioned whether there was any implications from using average prices, noting that for about half the regions prices bottomed in 2001. The only implication is a modest loss in precision relative to measures based on annual prices.

Thus, for the period in question UK producers appear to have suffered excess losses from Chilean supply expansion due to the feed quota of about 20%.

Tariff Effects

Results indicate that total welfare losses from the Chilean tariff ($155 million) are modest in relation to losses from the Norwegian tariff ($1.44 billion). Hence, discussion will focus on the Norwegian tariff unless indicated otherwise.

With the feed quota in place the lion's share of the total welfare loss, $1.30 billion, accrues to Norwegian producers (table 6). This should not be surprising given that 93% of the tariff's incidence is borne by Norwegian producers (recall table 4). The only beneficiaries of the tariff are Norway's international competitors, who, owing to the induced rise in their prices, enjoy a combined welfare gain of $138 million. Most of this gain, $72 million, accrues to Chilean producers, with UK benefitting by a more modest $24 million. The spillover benefits, however, are more than offset by consumer losses totaling $277 million. Hence, the tariff against imports from Norway exacts a high cost in relation to benefits.

Table 5. Welfare Effects of Increased Salmon Supplies from Chile on Salmon Importers and Exporters with and without Norway's Feed Quota, 2000-2002

Item	Welfare Gain		Ratio
	With Quota	Without Quota	
Exporters:	(Million US $)		
Norway	-381	-291	1.31
Chile	1,054	1,119	0.94
United Kingdom	-49	-40	1.20
Canada	-49	-39	1.24
Faroe Islands	-33	-27	1.23
ROW	-13	-11	1.21
All	529	711	0.74
Importers:			
European Union	399	314	1.27
United States	215	169	1.27
Japan	41	32	1.28
ROW	116	92	1.27
All	771	606	1.27
ΔTS	1,299	1,317	0.99

Lifting the feed quota reduces the total welfare loss to $1.21 billion and redistributes losses away from Norwegian producers to consumers (table 6). The most dramatic effect in terms of the research objectives of this paper is the enlargement of the implicit subsidy conferred by the tariff on Norway's international competitors. In particular, UK producer benefits from the tariff rise from $24 million to $79 million when the quota is removed, for a net gain of $55 million. Comparing this gain with the $9 million gain from quota removal estimated in table 5, tariff effects are more consequential than supply expansion effects. Adding the tariff-based gain of $9 million to the supply expansion-based gain of $55 million yields an estimated gross gain to UK producers from quota removal of $64 million. This is a *gross* gain because, according to table 6, feed quota removal would reduce the spillover benefit that UK producers receive from the Chilean tariff by $2 million. Taking this loss into account, the estimated net gain to UK producers from quota removal is $62 million, equivalent to 9.5% of UK export revenue over the evaluation period.

Table 6. Welfare Effects of U.S. Tariffs on Salmon Importers and Exporters with and without Norway's Feed Quota, 2000-2002

Item	Tariff Against Norway			Tariff Against Chile		
	With Quota	W/O Quota	Ratio	With Quota	W/O Quota	Ratio
Exporters:	(Mil $)			(Mil $)		
Norway	-1,298	-869	1.49	61	49	1.25
Chile	72	222	0.33	-116	-121	0.96
United Kingdom	24	79	0.31	9	7	1.26
Canada	21	62	0.33	8	6	1.26
Faroe Islands	15	45	0.33	6	5	1.26
ROW	6	21	0.31	2	2	1.26
All	-1,160	-441	2.63	-29	-52	0.56
Importers:						
European Union	-143	-393	0.36	-62	-49	1.26
United States	-78	-219	0.36	-46	-37	1.25
Japan	-14	-38	0.37	-7	-5	1.25
ROW	-42	-115	0.36	-12	-10	1.25
All	-277	-766	0.36	-126	-100	1.26
ΔTS	-1,436	-1,206	1.19	-155	-152	1.02

SENSITIVITY ANALYSIS

As Bredahl, Meyers, and Collins (1979) suggest, price transmission elasticities have important implications for trade analysis. Accordingly, since the classical trade model assumes transmission elasticities are unity, we re-computed our welfare measures with the nine transmission elasticities in the model set to one to determine how inferences might be affected. Results show the *magnitude* of the impacts being affected, but not their *direction* (Table 7).

For example, the total welfare losses from the Norwegian tariff decline from $1.44 billion to $1.02 billion when the price transmission elasticities are set to one. However, the overall conclusion that the feed quota increases the tariff's cost to Norwegian producers and reduces spillover benefits to Norway's international competitors is unaffected. Similarly, the total welfare gain of Chile's supply expansion increases from $1.30 billion to $1.46 billion when the price transmission elasticities are set to one. But the overall conclusion that the feed quota enlarges gains to consumers at the expense of Chile's international competitors is unaffected.

Table 7. Sensitivity of Welfare Effects to Price Transmission Elasticities

Item	Chile's Supply Expansion			US Tariff Against Norway		
	With Quota	W/O Quota	Ratio	With Quota	W/O Quota	Ratio
Exporters:	(Mil $)			(Mil $)		
Norway	-342	-268	1.28	-903	-655	1.38
Chile	1,113	1,162	0.96	39	119	0.33
United Kingdom	-47	-40	1.18	15	49	0.31
Canada	-70	-58	1.21	20	62	0.33
Faroe Islands	-27	-22	1.21	8	24	0.33
ROW	-12	-10	1.19	4	12	0.32
All	614	764	0.80	-817	-389	2.10
Importers:						
European Union	366	294	1.24	-89	-255	0.35
United States	291	234	1.24	-70	-202	0.35
Japan	65	51	1.26	-14	-39	0.36
ROW	129	104	1.24	-31	-89	0.35
All	851	684	1.24	-205	-584	0.35
ΔTS	1,465	1,448	1.01	-1,022	-973	1.05

Note: results are based on setting all price transmission elasticities in the model equal to one.

The simulations confirm that price wedges equal tariff rates when price transmission elasticities are unity (26.1% for Norway and 5.5% for Chile). In the original model the price wedges were 37.7% and 8.4% (see footnote 2), which explains why tariff impacts implied by that model are larger than in the sensitivity analysis. Still, our overall conclusion that the feed quota entails significant costs in terms of unintended consequences is not much affected by the transmission elasticities.

Concluding Comments

The basic theme of this research is that trade restrictions can have unintended consequences. In the case of the feed quota instituted by Norway to assist United Kingdom salmon producers, unintended consequences include steeper price declines associated with increased supplies of salmon from Chile, and a magnified Norwegian incidence of US tariffs on salmon imports from Norway. The latter undermined the tariffs' ability to benefit US producers, and decreased the implicit subsidies enjoyed by Norway's international competitors as a result of higher world prices induced by the tariffs. For UK producers, these unintended consequences resulted in some $62 million in foregone surplus between 2000 and 2002, equivalent to 9.5% of export revenue. Hence, the feed quota may well have been ineffectual in its main purpose, which was to benefit UK producers by restricting the supply of Norwegian salmon into the EU market.

This conclusion, coupled with the finding in this study that the general equilibrium demand curve for salmon is elastic at -1.2, suggests trade restrictions are not an effective policy instrument for assisting salmon producers. One policy alternative that may have merit is market promotion. Research suggests Chile is a major free rider on Norway's salmon promotion efforts (Kinnucan and Myrland 2003). Since these efforts are aimed chiefly at markets in the European Union and Japan, Chile could be encouraged to invest in promotion in the United States, perhaps in league with US producers. In addition to addressing the free-rider issue, the added promotion would help strengthen market prices, thereby moderating price declines associated with ongoing productivity improvements in salmon farming (Asche 1997; Anderson 2002). If Chile's promotion program were funded by an export tax, as is the case for Norway's program, the scheme would have the added benefit of reducing market supply. The point is that given the unintended consequences of supply control, demand-side policies might offer a better long-term solution to the problem of low or declining prices.

REFERENCES

Anderson, JL. "Aquaculture and the Future: Why Fisheries Economists Should Care." *Marine Resource Economics*, 17(2002), 133-51.

Asche, F. "Testing the Effect of an Anti-Dumping Duty: the US Salmon Market."*Empirical Economics*, 26(2001), 343-55.

Asche, F. "Trade Disputes and Productivity Gains: The Curse of Farmed Salmon Production."*Marine Resource Economics*, 12(1997), 67-73.

Asche, F; Bjørndal, T; Salvanes, KG. "The Demand for Salmon in the European Union: The Importance of Product Form and Origin." *Canadian Journal of Agricultural Economics*, 1998, 46, 69-81.

Asche, F; Bremnes, H; Wessells, C. "Product Aggregation, Market Integration, and Relationships Between Prices: An Application to World Salmon Markets." *American Journal of Agricultural Economics*, 1999, 81, 568-81.

Bendz, K. "Antidumping Duties on Imports of Salmon from Norway." Gain Report Number E35131. USDA, *Foreign Agricultural Service*, 2005, 1, July.

Bjørndal, T; Asche, F; Steen, F. "Price Determination for Fresh Salmon in the EU Market." *Bergen: Norwegian School of Economics and Business*. Unpublished manuscript, 1996.

Bjørndal, T; Salvanes, KG; Andreassen, JH. "The Demand for Salmon in France: The Effects of Marketing and Structural Change." *Applied Economics*, 1992, 24, 1027-34.

Bredahl, ME; Meyers, WH; Collins, KJ. "The Elasticity of Foreign Demand for U.S. Agricultural Products: The Importance of the Price Transmission Elasticity." *American Journal of Agricultural Economics*, 1979, 61, 58-63.

Buse, RR. "Total elasticities – a predictive device." *Journal of Farm Economics*, 1958, 40, 881-891.

Callendar McDowell. "Commentary: Shetland Salmon Farmers Commissioned Global Cost of Production Study." *Aquaculture Communications Group*. July 18, 2005.

http://www.aquacomgroup .com/index.cfm?fuseaction=main.dspReadImportNewsandID=608 (accessed 2 September 2005).

DeVoretz, DJ; Salvanese, KG. "Market Structure for Farmed Salmon." *American Journal of Agricultural Economics*, 1993, 75, 227-33.

Edelman, M; Langley, J; White, FC. *Agricultural Supply Control.* Staff Research Paper 523, Department of Economics, Iowa State University, 2003.

Fousekis, P; Revell, BJ. "Retail Fish Demand in Great Britain and its Fisheries Management Implications." *Marine Resource Economics*, 2004, 19, 495-510.

Herrmann, ML; Lin, BH. "The Demand and Supply of Norwegian Atlantic Salmon in the United States and European Community." *Canadian Journal of Agricultural Economics*, 1988, 38, 459-71.

Herrmann, ML; Mittelhammer, RC; Lin, BH. "Import Demand for Norwegian Farmed Atlantic Salmon and Wild Pacific Salmon in North America, Japan, and the EC." *Canadian Journal of Agricultural Economics*, 1993, 41, 111-25.

Hertel, TW; Tsigas, ME. "General Equilibrium Analysis of Supply Control in U.S. Agriculture." *European Review of Agricultural Economics*, 1991, 18, 167-91.

Houck, JP. *"Elements of Agricultural Trade Policies.* Prospect Heights." Illinois: Waveland Press, Inc, 1986.

Kinnucan, HW. "Futility of Targeted Fish Tariffs and an Alternative." *Marine Resource Economics*, 2003, 18, 211-24.

Kinnucan, HW; Myrland, Ø. "Free-Rider Effects of Generic Advertising: The Case of Salmon." *Agribusiness: An International Journal*, 2003, 19, 315-24.

Kinnucan, HW; Myrland, Ø. "Optimal Advertising Levies with Application to the Norway-EU Salmon Agreement." *European Review of Agricultural Economics*, 2000, 27, 39-57.

Kinnucan, HW; Myrland, Ø. "The Relative Impact of the Norway-EU Salmon Agreement: A Midterm Assessment." *Journal of Agricultural Economics*, 2002, 53, 195-219.

Kola, J. "Efficiency of Supply Control Programmes in Income Redistribution." *European Review of Agricultural Economics*, 1993, 20, 183-98.

Lem. A. "Salmon Situation Still Unclear." FAO Globefish. May, 2005.

Marsh, JM. "Impacts of Declining U.S. Retail Beef Demand on Farm-Level Beef Prices and Production." *American Journal of Agricultural Economics*, 2003, 85, 902-913.

Pearce, IF. Total demand curves and general equilibrium, *Review of Economic Studies*, 1952, 20, 216-27.

Piggott, RR. "Some Old Truths Revisited." *Australian Journal of Agricultural Economics*, 1992, 36, 117-40.

Steen, F; Asche, F; Salvanes, KG. "The Supply of Salmon in EU: A Norwegian Aggregated Supply Curve." *SNF Working Paper 53/97*. Centre for Fisheries Economics, Norwegian School of Economics and Business Administration. Bergen, Norway, 1997.

Steen, F; Salvanes, KG. "Testing for Market Power Using a Dynamic Oligopoly Model." *International Journal of Industrial Organization*, 1999, 17, 147-77.

Wessells, CR; Wilen, JE. "Economic Analysis of Japanese Household Demand for Salmon." *Journal of World Aquaculture Society*, 1993, 24, 361-78.

Wessells, CR; Wilen, JE. "Seasonal Patterns and Regional Preferences in Japanese Household Demand for Seafood." *Canadian Journal of Agricultural Economics*, 1994, 42, 87-103.

APPENDIX: SUPPLY SHIFT MEASUREMENT

Marsh (2003, p.902) defined shifts in the US retail demand curve for beef as the "percentage differences between observed retail beef prices and estimated retail beef prices holding demand constant." This definition can be adapted to the measurement of shifts in Chile's excess supply curve by reference to Figure 1.

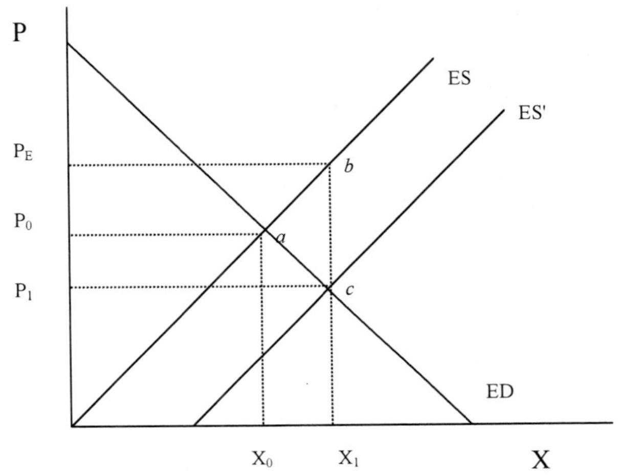

Figure 1. Measurement of the Vertical Shift in Chile's Excess Supply Curve

Let X_0 and P_0 represent Chile's observed exports and price in 2000 and let X_1 and P_1 represent the observed exports and price in 2002. The supply shift corresponding to these price/quantity coordinates is identified by letting P_E be the price that would obtain if the observed increase in exports was due to a demand shift rather than a supply shift. Specifically, in Marsh's terminology P_E is the estimated price holding supply constant, i.e., if demand increased along ES from point a to point b. Dropping a line from P_E to P_1 gives the absolute vertical shift in the excess supply curve, labeled in the diagram as distance $bc = |P_1 - P_E|$.

The corresponding relative vertical shift is:

$$\gamma = (P_1 - P_E)/P_E \qquad (A1)$$

where $\gamma < 0$ since the shift is downward.

In Marsh's (2003) paper a numerical estimate for γ was obtained using a four-step procedure. Here, we derive an analytical expression that permits the calculation to be done in one step.

From Figure 1, $P_E = P_1 + (P_0 - P_1) + (P_E - P_0)$, which implies:

$$P_E = P_0 + \Delta P \qquad (A2)$$

where $\Delta P = P_E - P_0$. Substituting equation (A2) into (A1) yields:

$$\gamma = (P_1 - P_0 - \Delta P)/(P_0 + \Delta P)$$

Multiplying the right-hand side of this expression by P_0/P_0 yields:

$$\gamma = (P^* - P_E^*)/(1 + P_E^*) \qquad (A3)$$

where $P^* = (P_1 - P_0)/P_0$ is the observed percentage change in price and $P_E^* = (P_E - P_0)/P_0$ is the percentage change in price measured along the original excess supply curve *ES* from point *a* to point *b*. Following Marsh (2003), P_E^* can be approximated as follows:

$$P_E^* \approx X^*/\varepsilon$$

where $X^* = (X_1 - X_0)/X_0$ is the observed percentage increase in exports and ε is the excess supply elasticity corresponding to *ES* . Substituting the above expression into equation (A3) yields:

$$\gamma \approx (\varepsilon P^* - X^*)/(\varepsilon + X^*), \qquad (A4)$$

which is identical to text equation (22).

In: Salmon: Biology, Nutrition and Consumption
Editors: P. Lacopo et al., pp. 187-199
ISBN: 978-1-60741-287-8

Chapter 8

"SALMON SOVEREIGNTY" AND THE DILEMMA OF INTENSIVE ATLANTIC SALMON AQUACULTURE DEVELOPMENT IN BRITISH COLUMBIA

John P. Volpe
Environmental Studies, University of Victoria, Canada

ABSTRACT

Agricultural industrialization and globalization are often identified as key processes in the erosion of food security and sovereignty. Global seafood production is dominated by modern day hunter-gatherers and so food sovereignty debates until recently have been confined to terrestrial systems. The recent introduction and subsequent spectacular proliferation of industrial marine aquaculture has brought the food sovereignty debate to the sea. The current hub of North American marine-based industrial aquaculture is British Columbia (BC) where salmon farming has grown to become a major factor in shaping coastal communities. Since its introduction into BC in the mid-1980s, the global industrial salmon farming industry has changed dramatically. The meteoric rise of farm salmon production exported from Norway, Chile, and Scotland has caused global salmon prices to plummet, forcing intense industry-wide rationalization. This is typically characterized by large multinationals cannibalizing smaller producers in order to establish economies of scale to counterbalance market price erosion of their product. As a result, global market forces have all but eliminated the sovereignty of both BC salmon farmers and salmon fishermen. Salmon farmers and fishermen—and their communities—must adopt ever-greater economies of scale to remain competitive with growing/consolidating foreign farm producers. All ecological issues associated with farms (organic wastes, parasites, toxins use, escapes etc.) are density dependant and therefore escalation of production results in incrementally greater ecological impacts. Further, the (d-)evolution of salmon from a culturally iconic seasonal delicacy and cultural keystone species of indigenous communities to a low-value global commodity has manifested a broad array of impacts highlighting the codependence of ecology, culture, and economics in coastal communities. Finally, increasing production increases supply, which in turn lowers price and alters consumer expectations of price/value relationships. This not only forces another round of production increases/consolidation on the farm, but also drives out commercial fishermen, particularly smaller owner/operators, who have no capacity to adopt economies of scale in the face of falling salmon prices. The ensuing rationalization process results in corporate fleets, the only model capable of adapting to the economic realities of the commoditization of salmon. Such loss of sovereignty effectively liquidates social capital of coastal communities which is instead invested into corporate machinery paying a lower rate of return far removed from the communities who are now left to pay off the interest as a legacy to the venture.

He who has bread has many problems. He who has no bread has only one problem.

—*Byzantine proverb*

INTRODUCTION

The recent surge of salmon farming along the coast of British Columbia represents an enormous restructuring of economy and environment that has had major implications for coastal communities. By way of setting context, we begin at *The World Forum on Food Sovereignty* held in Rome in June 2002 in the shadow of the much larger, production oriented, governments-only *World Food Summit. The Forum* provided a venue for citizens to extend discussions from simple production strategies, to encompass the rights of producers. Thus *The Forum* was a venue for the voice of the producer and put the concept of "food sovereignty" on the world stage; in retrospect it is now seen as a watershed event in how we address global food politics.

Why this event is viewed with such significance is revealed in its name which invokes "food sovereignty" instead of the more familiar "food security." More than 200 definitions of "food security" could be found in the literature just over a decade ago (Maxwell and Smith 1992). In all cases the right to access and consume food is emphasized. Food sovereignty balances the rights of people to consume healthful food with protecting the autonomy of [often oppressed or marginalized] food producing groups and their rights to decide how to produce food and how to sustain their environments and livelihoods. Food producers are in fact the majority of the world's population and, ironically, the overwhelming majority of the world's hungry (Roberts 2003). Numerous examples exist where the rights of producers and consumers have become unbalanced, particularly where states are obliged to feed all citizens but lack the means to develop and protect the capacity to grow food. For instance, during the Irish potato famine of 1845–46, the amount of food Ireland exported to Britain would have been sufficient to feed those who starved. The 1974 Bangladesh famine resulted from people displaced from employment by floods, leading to widespread economic breakdown—even though more food was produced in the country that year than those years immediately before or after (Sen 1981). These cases illustrate a more general phenomenon; hunger is often a result of inadequate access to, not lack of food.

This is not to suggest the supply of food has always exceeded demand, nor that it always will. In the middle of the last century many populations around the world were certainly food deficient. However this had little apparent effect in stemming population increase. The situation became acute when food demands rose sharply in the baby boom years following World War II when most arable land was already under culture. An array of technical advances (artificial fertilizers, pest-/herbicides, irrigation, artificial selection) combined to produce high yield crop strains genetically capable of responding to the new production model of intense agrichemical and irrigation input. These, combined with adoption of increased mechanization and energy use on the farm, resulted in unprecedented increases of production per acre. When viewed together these technologies summed to enable the large-scale conversion of oil and water into grain. This so called *Green Revolution* ushered in the age of industrial agriculture which succeeded in meeting the demands of the day and has supported the resultant growth of the human population to the present even though the number of farmers working the land in the developed world continues to rapidly decline.

Production capacity of these now conventional technologies, however, is being exhausted and history is poised to repeat as the global human population finds itself on the brink of food deficiency once again. Production per acre has plateaued and in many cases declined, reflecting degradation of the land's production capacity by Green Revolution industrialization (Tilman et al. 2002). The agribusiness funded scientific community has predictably responded by suggesting a new generation of advances—e.g. GMOs (genetically modified organisms)—must be implemented if widespread famine is to be avoided (U.S. National Academy of Sciences 2000). What goes unreported in such techno-centric analyses is that presently global agriculture produces more than adequate supplies to feed every person on Earth a 2,500-calorie daily ration. Despite this, 60 million people die annually as a direct result of insufficient access to food. To put a finer point on the troubling ethical corollary this raises; the same number of people could be adequately fed, thus ending the crisis, by the grain saved if Americans reduced their intake of meat by 10% (Schwartz 2001). Today, as in the past, the critical issue for much of the world's poor remains access to otherwise abundant food, not a shortfall of production. Nonetheless, the complex and nuanced drivers responsible for decreasing global food security and sovereignty are often ignored, resulting in simplistic trickle down prescriptions: hungry people will be fed if we can just produce more food *ipso facto* GMOs are the silver bullet.

Along with assorted techno-fixes such as GMOs, aquaculture—the farming of marine and freshwater species—is now being promoted as a potential solution to meet the perceived shortfall in demand for protein. Aquaculture has an ancient history beginning in China during the Zhou Dynasty, approximately 2,300 years ago (Li 1994). Within 200 years (Han Dynasty) the practice of polyculture (simultaneous culture of multiple, synergistic species) had become widely established and remains a fixture to this day in China and across much of the world. In contrast, modern industrial aquaculture is characterized as being reliant on external inputs of energy towards the production of an export crop with a primary objective of generating profit not food. However, the potential protein production capacity of aquaculture cannot be ignored and many are now calling for a *Blue Revolution*—application of new technologies in order to repeat, in the world's oceans, the "successes" of the Green Revolution on land. Proponents cite three major objectives of the Blue Revolution: to increase food security for those currently or soon to be in deficit; to relieve harvest pressure on wild stocks; and to provide economic opportunity for struggling communities (Skladany and Harris 1995). Deployment of the Blue Revolution in the developed world is based exclusively on intensive production models favouring salmon and shrimp operations over alternative, more sustainable lower trophic level species and more balanced production models such as polyculture (Skladany and Harris 1995).

The Blue Revolution has clearly begun, but it differs from its predecessor, the Green Revolution, in two important ways. First, the Green Revolution's success was based on innovations that enabled increased production without increasing the spatial footprint of operation. While some increases in efficiency in the Blue Revolution are evident, the majority of its increasing production is due to simply increasing the physical area used for operations. Second, the Green Revolution focused on grain (primarily carbohydrate) production, the backbone of the human diet. In contrast, protein is the product of the Blue Revolution. Protein is obviously an important dietary component, however it is very much secondary to carbohydrates in terms of feeding food deficient populations.

The dominant case history that frames the development of the Blue Revolution is salmon farming. Begun in Norway in the 1970s, farmed (Atlantic) salmon has grown to be a dominant global aquaculture product. This process was not without its setbacks. After a decade of meteoric growth, the effects of the industry began to be seen in Norwegian wild salmon populations. The detrimental effects of genetic introgression via farm-wild hybridization have been well documented (Fleming et al. 2000). However the most dramatic effects have been those involving disease and parasite transfers among wild and farmed fish. One of the most alarming events occurred in the early 1980s when farm-borne epidemics of the potentially fatal parasite *Gyrodactylus* devastated wild fish populations. Fuelled by public outrage, the ensuing Norwegian government crackdown on industry resulted in its major reconfiguration. Many operators adapted to the regulations, while others began looking for alternative venues for their operations.

Concurrently, in September 1984 Brian Mulroney was elected Prime Minister of Canada. High on his agenda was the replacement of the Foreign Investment Review Act with the Investment Canada Act. The new legislation essentially removed the necessity that Canadians citizens hold majority ownership of Canadian registered companies.

And so the stage was set for the migration of Norwegian salmon farm companies to Canada. At home in Europe, Norwegian companies were being compelled to conform to strict and in some cases costly new operating procedures, whereas the Canadian federal government threw open the door for them to operate unfettered in Canada. The British Columbia coastline provided exemplary physical and biological habitat required by industry, and regulators were positively predisposed to the arrival of the Norwegians to help solidify the fledgling BC salmon farming industry, at that time based exclusively on farming Pacific (*Oncorhynchus*) salmon species. The influx of Norwegian operators also meant the importation of the Norwegian species of choice—Atlantic salmon (*Salmo salar*).

> We are very strict about the quality and the environmental questions [in Norway]. Therefore, some of the fish farmers went to Canada. They said, 'We want bigger fish farms; we can do anything; we can do as we like.'
>
> —*Norwegian Parliamentarian. Transcript of the House of Commons Standing Committee on the Environment 12 September 1990*

The Norwegians by this time had spent unprecedented resources establishing an international export market for its farmed Atlantic salmon. It was not about to undermine that success by introducing a novel (in the eyes of their major markets at that time) product in the form of farmed Pacific salmon. The transition in British Columbia's aquaculture industry from Pacific salmon to Atlantic salmon was swift (Figure 5.1).

Currently (as of 2006) Atlantic salmon comprises 76% of farmed salmon in BC (chinook salmon and some coho comprising the balance, both Pacific salmon species). Not only did the species change with the influx of Norwegian salmon farms, so did production volume. In 1988, during the early days of the industrialization process, BC farmers produced only 80,000 kilograms of Atlantic salmon; in 2002, with the industry in full blush, production exceeded 71,600 metric tonnes, representing an interim production increase of 895,000 times (Table 5.1). Today BC is the fourth largest producer of Atlantic salmon in the world, behind Chile, Norway, and Scotland respectively. While BC's industry has shown impressive growth over the past two decades, industry growth in competing countries was even greater, lead by Chile which in 1990 did not produce a single Atlantic salmon, but is currently the world leading Atlantic salmon grower, with 280,086 metric tonnes produced in 2003.

Table 5.1. BC wild salmon and farm Atlantic salmon production and value, contrasting changes from 1988 when Atlantic salmon aquaculture was initiated to the industry's peak performance period in 2002

Species	1988 Production (x1000 mt)	2002 Production (x1000 mt)	% Change	1988 Ex-vessel (CAD$ / kg)	2002 Ex-vessel (CAD$ / kg)	% Change
Chinook	5.92	1.69	-71	7.40	4.19	-43
Chum	30.30	12.35	-59	2.73	0.49	-82
Coho	7.08	0.47	-93	5.33	2.34	-56
Pink	32.22	8.61	-73	1.53	0.33	-78
Sockeye	11.94	10.15	-15	8.07	3.97	-51
	1988 Production (*x1000 mt*)	2002 Production (*x1000 mt*)	Change	1988 Farm Gate (CAD$ / kg)	2002 Farm Gate (CAD$ / kg)	% Change
Farm Atlantic	0.08	71,600	+895,000x	7.40	3.33	-55

Source: Fisheries and Oceans Canada

There is little doubt the tremendous growth of farm salmon production over this period played a significant role in the value erosion of wild product.

The massive global increase in farmed salmon production resulted in falling wholesale salmon prices, for both farmed *and* wild salmon. The immediate consumer benefit was clear; salmon had become affordable to a much larger proportion of the population. However, from the perspective of producers, particularly those in BC whose business plans had been developed in the heyday of salmon prices, the new reality of a globalized salmon commodity market has represented a significant challenge to fiscal viability.

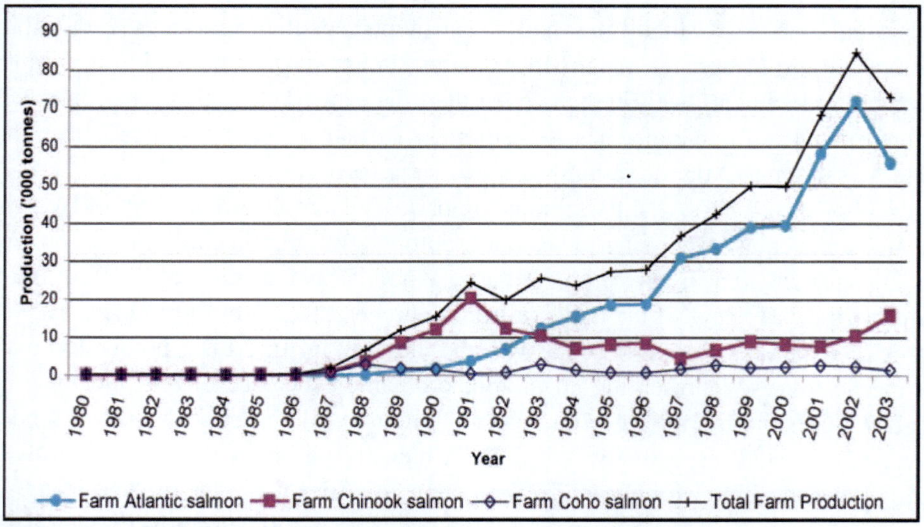

Source: Fisheries and Oceans Canada

Figure 5.1. Farm salmon production in BC by species 1980–2003. The first record of commercial salmon aquaculture in BC was in 1980 using coho salmon

From 1988 to 2002 the farm-gate value of BC farm salmon dropped 55%. In order to overcome this precipitous decline, producers in BC as elsewhere, were forced to adopt ever-increasing economies of scale, the strategy being that in order to preserve a stable profit margin in a market of diminishing unit value, one must produce more units and do so as cheaply as possible. Production in 2002 increased 895,000 times over that of 1988. While the unit value of farm salmon had declined dramatically (55%); this was more than compensated for by the six orders of magnitude production increase. BC is by no means unique. The same scenario has been played out in all salmon farming nations, with each company in competition with all others to produce the most salmon for the least cost. The competitive profile of the industry is complicated by the fact that the vast majority of global production is generated by relatively few multinational companies, each operating in many countries and so pitting one region against another. It is this economic reality from which most issues currently associated with industrial aquaculture arise; the globalization of salmon has transformed this fish from a seasonal, high value delicacy to a low value commodity available year-round that comes with significant environmental—and thus public relations implications.

Discussion

What are the implications of this situation for coastal communities and their food security? The high profile issue of the predominance of sea lice on salmon farms and the resultant increased infection of juvenile wild salmon is an example of how the globalization of salmon production can generate serious localized environmental impacts. In BC the term "sea lice" refers to either of two ectoparasitic crustaceans, *Lepeophtheirus salmonis* and *Caligus clemensi*. Both species naturally occur in coastal BC waters and thus are natural components of the marine ecosystem. Both are known to infect wild and farmed salmon in

BC. The former is a salmon specialist and the latter a generalist, known to infect numerous non-salmonid fish species as well as salmon. The lifecycle of both types of sea lice is generally the same: fertilized eggs (up to many hundred per female) are attached to the female who is in turn attached to an adult salmon in the ocean. In time the eggs hatch into free-swimming nauplii larvae which in time grow and advance through copepodid, çhalimus, and finally adult stages. Relatively early in this progression the young sea louse must find a host salmon (or perhaps one of a short list of other species in the case of *C. clemensi*). Given that the louse is free floating in the Pacific Ocean, the chances of literally running into a prospective host (not to mention successfully attaching) are extraordinarily slim. Therefore it is no surprise that natural sea lice abundances tend to be quite low. This scenario changes dramatically when salmon farms enter the equation. Salmon farms, by virtue of being home to as many as 1.5 million salmon in very small area, provide a high chance of success for any nauplii lucky enough to float through. Eventually such nauplii attached to their host mature and reproduce nauplii of their own. However, now the chances of successfully finding a host are very, very good—resulting in more farm salmon infections—and the cycle repeats itself until, if left unchecked, epidemic conditions rapidly set in. While this is clearly bad news for the farm salmon, how does this situation affect wild salmon?

We start by asking where are salmon farms situated? They are not randomly dispersed along the coast, but are placed in areas that, among other things, provide shelter from the open ocean (i.e. storms etc.). For this reason the head of bays and fjords are particularly attractive; but this is where the outlets of salmon bearing rivers are also found. In the spring and early summer when young wild salmon smolts migrate from their home rivers to the ocean they do so in the absence of adult salmon which are feeding offshore. Thus, under natural conditions, the absence of hosts in the inshore areas keeps juvenile lice abundance very low, allowing juvenile salmon to pass through with little impact. However, the situation can change dramatically if a salmon farm is located on the smolts' migration path. Research in the Broughton Archipelago has shown that an average farm can amplify local sea lice abundance 33,000 times over ambient levels, resulting in an infection rate of over seventy times what would normally occur. Further, this highly magnified farm "lice footprint" was detectable thirty kilometres from the salmon farm under investigation (Krkošek et al. 2005), considerably beyond the 'organic footprint' (Parrish et al., Chapter 3, this volume). The lice are spread even further as infected juvenile salmon continue on their migration out to sea. Most will succumb to their infection. However they may cover considerable distances before they do and all the while the once juvenile sea lice attached to and feeding on them grow and mature into reproductive adults. By the time these sea lice are ready to reproduce they may have traveled twenty or thirty kilometres from their originating farm. In the interim, new juvenile salmon, having not been exposed to a salmon farm, have joined the migrating school from other rivers along the migration corridor. But the now mature sea lice, having been picked up as nauplii thirty kilometres away, are now producing their own eggs and nauplii, and those young smolts recently joining the migration become infected. Krkošek *et al.* (1995) documented this phenomenon and witnessed how, through this process, a single farm infected otherwise unexposed salmon smolts more than thirty kilometres distant. These fish, in turn, extended the original farm's sea lice footprint to over seventy kilometres (the limit of the research survey). Dynamics of this nature are widely accepted as being responsible for spectacular collapses of Broughton Archipelago pink and chum salmon runs (Krkošek et al. 2005; Morton et al. 2004, 2005).

These data are no surprise to any one familiar with (1) fundamental epidemiology, or (2) the history of salmon aquaculture. It is very well understood that clustering hosts into exceptionally high densities and exposing them to contagion invites epidemics. Even if this theoretical expectation is not compelling, we need only look at the experiences of other jurisdictions that have gone before. Indeed, a very similar series of events occurring in Norway previously, and referred to above, precipitated the rapid expansion of salmon aquaculture in BC. Thus, we should not be surprised to see history repeating itself, particularly given that many of the founders of BC's current industry are those who resisted Norwegian government imposed reform in the aftermath of the *Gyrodactylus* incidents there.

Many have asked: "If these events were predictable, then why were they not avoided, or at least remedied following identification?" On the surface the answer appears complex and includes, among others, multi-scale ecological dynamics and aggressive political agendas. However, the actual underlying explanation is much more simple—sea lice epidemics (along with other environmental issues around salmon net pens) are in fact fated to occur and reoccur as a result of the globalization of the salmon industry.

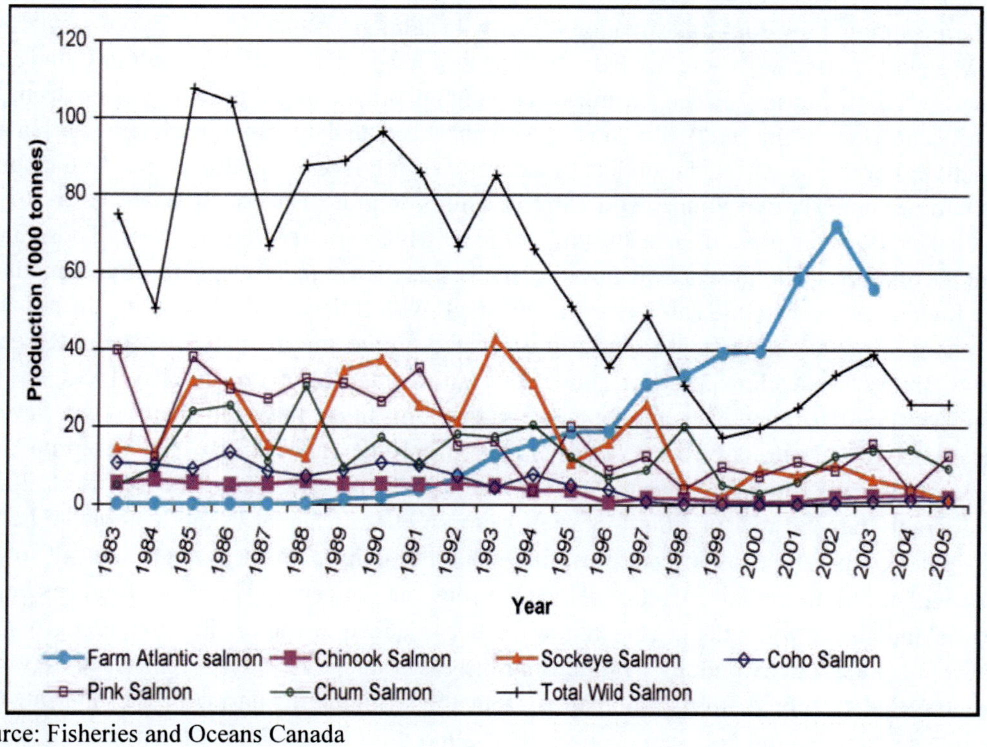

Source: Fisheries and Oceans Canada

Figure 5.2. Production of the BC commercial salmon capture fishery 1983–2005 by species plotted with production of BC farm Atlantic salmon. Data available to 2003 only.

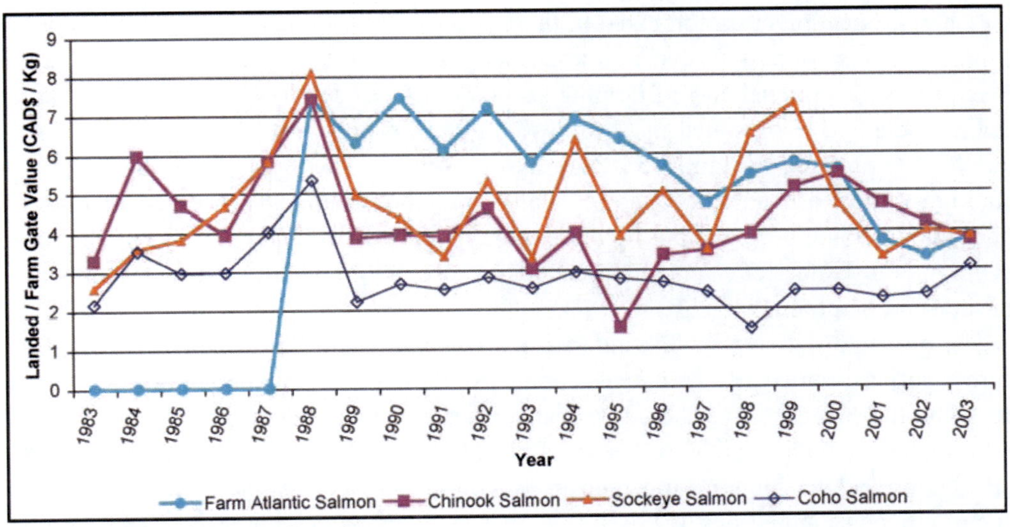

Source: Fisheries and Oceans Canada

Figure 5.3. Landed value of the top three most valuable commercially caught species plotted with farm gate value of farm Atlantic salmon 1983–2003

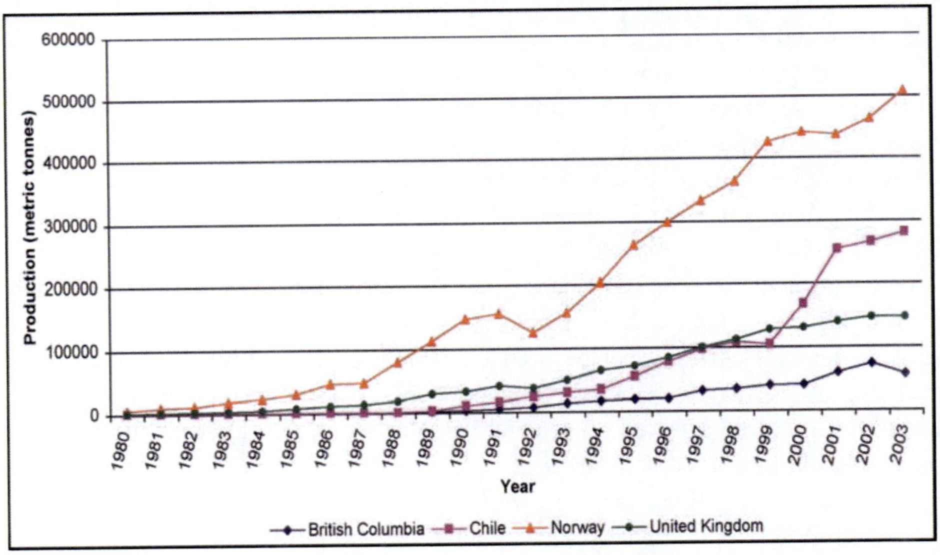

Source: UNFAO

Figure 5.4. Global production leaders of farm Atlantic salmon 1980–2003

The most desired salmon farming areas on the BC coast are in waters where the adjacent land is wilderness (indeed this is typical of much of the coast). This is key, because it means there is no road access proximate to farms; all materials coming and going (feed, crew, product, waste, equipment, etc.) are conveyed by marine transport. This is exceptionally expensive relative to regular truck transport—which, thanks to the developed coastlines of Norway, Scotland, much of Chile, etc., is used to much greater degree in these regions. The reliance of the BC industry on marine transport rather than trucks represents a significant increase in production costs compared to those incurred by the international competitors, but BC producers cannot pass this cost on to consumers without losing market share. Therefore, this increases the nominal cost of farming salmon in BC. The rapid escalation of global farm salmon production has resulted in a worldwide glut of supply, ever diminishing the product value for BC producers (Figures 5.2–5.4) and resulting in a wide scale transition of salmon from high value seasonal delicacy to low value common commodity. The new reality of the global salmon commodity market tightly constrains options for adding product value in BC, leaving the only hope for maintaining economic solvency in adopting greater economies of scale and cutting production costs.

The obvious first step in fighting sea lice epidemics in wild salmon is to adopt strict tenure sighting protocols to minimize cumulative farm effects including sea lice on wild stocks. In other words, farms should be spatially separated—as adopted in Norway following the *Gyrodactylus* outbreaks there. The incremental cost to Norwegian producers was modest. In BC, however, the salmon farm industry has reacted in exactly the opposite fashion. As value of farmed salmon drops, the need to cluster farms more and more tightly increases so as to minimize travel (i.e. cost) between tenures during the course of operations. This has resulted in the majority of BC salmon farm sites being highly clustered, with each cluster being located as close as possible to a town serving as a transportation hub. Major clusters are currently situated around Vancouver Island adjacent to Port Hardy, Port McNeil, Campbell River, and Tofino (Figure 5.5). Proposed industry expansion north will see Prince Rupert on the mainland added to the list.

The economic reality constraining the BC salmon farm industry is undeniable. The obvious solution to increased sea lice infestations would be to reduce farm densities, to move farms away from wild salmon smolt migration corridors and/or to fallow farms during the spring out-migration. Taking these actions, however, would not be economical.

This clash between environmental responsibility and economic viability represents a general pattern that is seen repeatedly across the array of issues associated with industrial scale salmon aquaculture. For instance, as value of farmed salmon erodes, the nominal *per capita* cost of an Atlantic salmon escapee diminishes too. Incentive for costly net maintenance declines as leaky net pens become more cost effective—revenue lost through escapes becomes less than the labour and material costs consumed to retain them.

Social dynamics are affected by product value erosion too. Labour is the second largest operating expenditure for salmon farms after feed. Falling salmon prices necessitate greater automation of the production system to trim costs. From 1997 to 2003 BC farmed salmon production increased approximately 100%, but on a per-unit production basis, wages, and salaries dropped 29% across the BC aquaculture sector (Cox 2005). Diminishing returns to coastal residents removes freedom to innovate and explore alternative production models leaving only intensifying industrialization as an option. In other terms, food sovereignty is forfeited and options are lost leaving communities with little choice but further investment in an industrial model with a bleak future.

Value erosion as a consequence of farm over production also sets in motion an analogous depensatory process in the salmon capture fishery. From 1988 to 2002 the production of every salmon species in the BC capture sector declined between 15% and 71%. Contrary to the basic tenets of economic supply and demand, as the supply decreased so did its value, up to 82% as in the case of chum salmon (*Oncorhynchus keta*) (Table 5.1). A simple analysis of catch values suggests fishermen today must harvest at a rate of up to five times that of 1988 to achieve an equivalent earning power. This is a conservative estimate owing to the fact that the 1988 figures have not been adjusted for inflation. Thus as prices drop, fishing pressure must increase to preserve earning power. Increased fishing pressure leads to smaller stocks which leads to increasing pressure and the depensatory cycle is in motion—greatly amplified by the negative market effects of an over abundance of farmed salmon.

On the surface, a multinational salmon farming conglomerate and an Amazonian rancher do not have much in common. However both have little sovereignty in how they make their living. The contexts are dramatically different; the underlying driving processes are the same. Both are locked in a battle, not against competitors for market share or restrictive government policy, but against a kind of mass psychosis so pervasive as to be iconic of modern societies: the myth of cheap. "Cheap" salmon or "cheap" beef in the case of the Amazonian rancher is an abstraction. In fact, "cheap" does not exist. Costs are transferred at various points in the production chain, but they do not disappear. The most damaging aspect of the industrial salmon farming experience is the establishment of a consumer culture expecting that salmon should be available fresh year round for the same price as chicken. And while this can be done, it is accomplished through massive economies of scale that demand significant ecological and social subsidies. These subsidies are not reflected in the retail price, but are costs just the same. Consumption of clean oxygenated water, removal of organic, and other wastes by currents, assimilation of escaped fish and greatly amplified numbers of sea lice in the ecosystem are just the start of the real costs transferred to the environment and to society at large.

Salmon farmers, like everyone else, would like to see the need for these subsidies eliminated. However, the reality is that farmed salmon is now a low value global commodity and that BC, relative to Norway or Chile, is an expensive place to farm salmon. The options for this dilemma are to either stay the course and pay the escalating social and environmental costs, or to abandon the industrial production model in favour of producing lower volume / higher value niche products (e.g. organic or "green" salmon). Both federal and provincial governments are clearly in favour of staying the course. The strategy seems to rest in the faulty belief that an as yet unattained "super-scale" of production will lead to sustainability.

Currently, as of 2006, we are weathering another major round of salmon farm industry expansion, extending north along the coast to Prince Rupert, in an effort to counter and compete with the rampant growth in Chile's farm salmon sector, which is showing no signs of slowing. The value of both farmed and wild salmon will undoubtedly continue on their conjoined journey of devaluing and decline, wreaking havoc enroute. The time for innovative development appears to be behind us now and those in the BC salmon sector—both farmed and wild—are little more the spectators, robbed of their sovereignty by shortsighted government policy driven by unrealistic consumer expectation. Ultimately, the loss of food sovereignty will translate to loss of food security, and this needs to be recognized by all those responsible for the salmon.

CONCLUSION

Three key points emerge from this analysis:

(i) Ecological issues that so often define the salmon farming controversy in BC and elsewhere, are in fact physical manifestations of underlying socioeconomic drivers. Working to solve the ecological problems while ignoring underlying drivers amounts to treating the symptoms while ignoring the disease. Efforts to conserve profitability in a globalized, commodity-based market (e.g. by increasing economy of scale) have resulted in simply transferring costs of production to associated ecological and social systems.

(ii) The fates of both BC salmon farmers and fishers are not in their own hands but are instead tied to global salmon market trends—both groups have seen their sovereignty eroded to the point of losing all control. Each, in their own way, is locked into reactionary strategies aimed at surviving the next round of incremental farm production increases without being able to plan for long-term sustainability.

(iii) Salmon farms in BC are forced to operate in what can be considered "hostile" territory. The physical realities of the BC coast impose production costs on BC operators that are not borne by their international competitors. Further, the social/cultural climate of coastal BC is increasingly unfriendly, resulting in further escalations in the cost of doing business here—again a burden not borne by foreign competitors. These factors, together with the seemingly limitless potential for expansion in other jurisdictions, particularly Chile, result in the future of industrial salmon farming in BC looking very bleak. Furthermore, industrial salmon farming is weakening wild salmon production due, for example, to the degrading effects of escalating sea lice populations and escapement of Atlantic farmed salmon into Pacific salmon habitat. If these effects are not remedied soon, both salmon sectors are threatened. For the First Nations and others who rely on wild salmon for cultural reasons as well as nutritional, the prospects look even more grim.

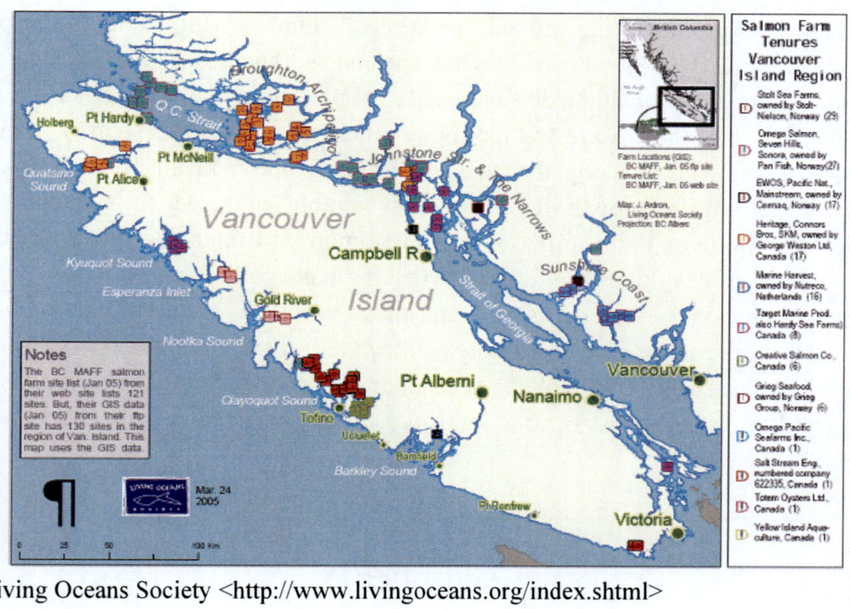

Source: Living Oceans Society <http://www.livingoceans.org/index.shtml>

Figure 5.5. Salmon farm tenure locations in BC. Note clustering of farms near transportation access points around Vancouver Island

After consideration, one sees that sovereignty within the BC salmon industry, from perspective of both fish farmers and fishers of wild salmon, has been dramatically compromised and in all likelihood is beyond recovery so long as current market trends continue and consumers fail to discriminate between the full costs of industrial and sustainable salmon products.

INDEX